Sustainable Textiles: Production, Processing, Manufacturing & Chemistry

Series Editor

Subramanian Senthilkannan Muthu, Chief Sustainability Officer
Green Story Inc.
Tung Chung, Hong Kong

This series aims to address all issues related to sustainability through the lifecycles of textiles from manufacturing to consumer behavior through sustainable disposal. Potential topics include but are not limited to: Environmental Footprints of Textile manufacturing; Environmental Life Cycle Assessment of Textile production; Environmental impact models of Textiles and Clothing Supply Chain; Clothing Supply Chain Sustainability; Carbon, energy and water footprints of textile products and in the clothing manufacturing chain; Functional life and reusability of textile products; Biodegradable textile products and the assessment of biodegradability; Waste management in textile industry; Pollution abatement in textile sector; Recycled textile materials and the evaluation of recycling; Consumer behavior in Sustainable Textiles; Eco-design in Clothing & Apparels; Sustainable polymers & fibers in Textiles; Sustainable waste water treatments in Textile manufacturing; Sustainable Textile Chemicals in Textile manufacturing. Innovative fibres, processes, methods and technologies for Sustainable textiles; Development of sustainable, eco-friendly textile products and processes; Environmental standards for textile industry; Modelling of environmental impacts of textile products; Green Chemistry, clean technology and their applications to textiles and clothing sector; Eco-production of Apparels, Energy and Water Efficient textiles. Sustainable Smart textiles & polymers, Sustainable Nano fibers and Textiles; Sustainable Innovations in Textile Chemistry & Manufacturing; Circular Economy, Advances in Sustainable Textiles Manufacturing; Sustainable Luxury & Craftsmanship; Zero Waste Textiles.

Subramanian Senthilkannan Muthu
Editor

Renewable Energy Implications for the Textile Industry

Springer

Editor
Subramanian Senthilkannan Muthu
Green Story Inc.
Tung Chung, Hong Kong

ISSN 2662-7108 ISSN 2662-7116 (electronic)
Sustainable Textiles: Production, Processing, Manufacturing & Chemistry
ISBN 978-3-032-10916-3 ISBN 978-3-032-10917-0 (eBook)
https://doi.org/10.1007/978-3-032-10917-0

This Springer imprint is published by the registered company Springer Nature Switzerland AG
The registered company address is: Gewerbestrasse 11, 6330 Cham, Switzerland

This book is dedicated to:
The lotus feet of my beloved Lord
Pazhaniandavar
My beloved late Father
My beloved Mother
My beloved Wife Karpagam and
Daughters Anu and Karthika
My beloved Brother
Last but not least
To everyone working in various industrial
spaces to make the world "SUSTAINABLE"

Contents

Introduction to the Book: Renewable Energy Implications for the Textile Industry

Subramanian Senthilkannan Muthu

Abstract Renewable energy and its implications in textile production represent a vital step toward transforming one of the most resource-intensive and environmentally impactful industries into a more sustainable and resilient sector. The textile industry, with its high dependence on electricity, thermal energy, water, and chemicals, has long been associated with significant greenhouse gas emissions and environmental burdens, but the integration of renewable energy sources such as solar, wind, biomass, and small-scale hydropower offers an opportunity to break this trend. By reducing reliance on fossil fuels and shifting to clean energy alternatives, textile units can not only cut their carbon footprint but also lower operating costs, stabilize energy supply in the face of rising fuel prices, and strengthen compliance with global sustainability standards demanded by brands and consumers alike. Furthermore, renewable energy adoption in textiles goes beyond environmental responsibility; it enhances brand reputation, creates competitive advantages in eco-conscious markets, and supports international climate commitments while contributing to local energy security and job creation. As the world moves toward greener supply chains and stricter sustainability certifications, the role of renewable energy in textile manufacturing is no longer a distant option but an immediate necessity, shaping the industry's future toward efficiency, innovation, and climate resilience.
Renewable energy and its implications in textile production represent a vital step toward transforming one of the most resource-intensive and environmentally impactful industries into a more sustainable and resilient sector. The textile industry, with its high dependence on electricity, thermal energy, water, and chemicals, has long been associated with significant greenhouse gas emissions and environmental burdens, but the integration of renewable energy sources such as solar, wind, biomass, and small-scale hydropower offers an opportunity to break this trend. By reducing reliance on fossil fuels and shifting to clean energy alternatives, textile units can not only cut their carbon footprint but also lower operating costs, stabilize energy supply in the face of rising fuel prices, and strengthen compliance with

S. S. Muthu (✉)
Green Story Inc., Tung Chung, Hong Kong

S. S. Muthu (ed.), *Renewable Energy Implications for the Textile Industry*, Sustainable Textiles: Production, Processing, Manufacturing & Chemistry,
https://doi.org/10.1007/978-3-032-10917-0_1

global sustainability standards demanded by brands and consumers alike. Furthermore, renewable energy adoption in textiles goes beyond environmental responsibility; it enhances brand reputation, creates competitive advantages in eco-conscious markets, and supports international climate commitments while contributing to local energy security and job creation. As the world moves toward greener supply chains and stricter sustainability certifications, the role of renewable energy in textile manufacturing is no longer a distant option but an immediate necessity, shaping the industry's future toward efficiency, innovation, and climate resilience.

The second chapter showcases the insights from a case study of renewable energy and environmental sustainability which are levers for textile companies' sustainable marketing activities. Focusing on the textile sector, this chapter presents the single case study of a medium-sized Italian textile and clothing company. This chapter highlights how environmental sustainability, proactively implemented by the company primarily through the adoption of renewable energy, can generate what we originally define as a "positive domino effect". The company's sustainability push, being absolutely crucial for counteracting the negative and harmful impacts of the textile industry, initially affects the environmental sustainability domain, which, in turn, creates a positive cascading effect on social and economic sustainability, thus holistically encompassing all sustainable marketing activities. This chapter reveals that the "positive domino effect" is achievable due to the textile company's efforts in environmental sustainability and renewable energy, which have enabled it to strengthen its market reputation while concurrently gaining the trust of numerous involved stakeholders. The company serves as a case in point in this chapter, demonstrating its emergence as a major contributor to sustainability initiatives within the Italian textile landscape, alongside its significant contributions to the local community and the industrial textile district in which it is located. This chapter offers significant theoretical implications, relevant practical indications for managers and policymakers, and avenues for future research.

The third chapter deals with the renewable energy and its implications in textile. The textile industry is a major consumer of water and energy, with substantial reliance on fossil fuels, resulting in significant environmental challenges. The finite availability and unsustainable nature of fossil fuels, combined with their contribution to environmental degradation, underscore the urgency of transitioning to alternative energy sources. This chapter examines the energy-intensive operations of the textile industry, including fiber production, yarn manufacturing, dyeing, finishing, and waste management, and explores the integration of renewable energy solutions such as solar, wind, biomass, and geothermal technologies. Through a comprehensive analysis, the chapter also highlights case studies and best practices that demonstrate the feasibility and benefits of adopting renewable energy. Key focus areas include reducing carbon emissions, lowering operational costs, and enhancing environmental sustainability. Economic and technological considerations, including cost-effectiveness, scalability, and compatibility with existing systems, are also discussed to address challenges in the transition process. The insights aim to provide a roadmap for industry stakeholders to adopt sustainable energy practices, paving the way for a greener and more resilient textile production model. The insights provided aim to guide industry stakeholders toward a greener, more sustainable textile production model.

The fourth chapter is about the cotton stalk bioenergy which is a renewable approach for sustainable textile manufacturing. Cotton stalks, a by-product of cotton cultivation, are often left on the fields or burnt, contributing to environmental pollution and waste. This chapter examines the potential of cotton stalks as a renewable bioenergy source for sustainable textile production. With their high lignocellulosic content, cotton stalks can be efficiently converted into biofuels through processes such as pyrolysis, gasification, and anaerobic digestion. This biomass valorization not only provides a decentralized and cost-effective energy supply for rural textile operations but also addresses waste management and soil degradation concerns. This chapter details the physicochemical properties of cotton stalks that make them suitable for energy production and explores sustainable technologies for their conversion. Integrating cotton stalk bioenergy into textile value chains reduces dependency on fossil fuels and promotes circular economy principles. Additionally, socioeconomic advantages like increased farmer income and rural electricity availability are examined alongside environmental advantages like reduced greenhouse gas emissions and better air quality. Successful implementation strategies are demonstrated by case studies from India and other cotton-producing countries, while techno-economic evaluations offer information on viability and the return on investment. Policy recommendations include subsidies for biomass energy projects, research support for advanced bioenergy systems, and awareness programs for farmers and textile stakeholders. This chapter concludes that leveraging cotton stalks for bioenergy not only strengthens the sustainability of the textile sector but also contributes meaningfully to climate resilience and rural development.

The fifth chapter deals with the renewable energy in textiles in addressing climate change. The textile sector is a major contributor to global greenhouse gas emissions because of its high dependence on fossil fuels for energy-intensive processes like dyeing, spinning, and finishing. In the wake of climate change, incorporating renewable energy sources into textile manufacturing offers a sustainable and economically feasible solution. This chapter explores the environmental footprint of conventional textile energy sources and highlights the shift towards solar, wind, geothermal, and biomass energy in textile production. The implementation of renewable energy along with reducing the carbon footprint also enhances operational efficiency and enduring productivity. Technological developments, like energy-efficient machinery and green chemistry, are discussed in the context of energy demand reduction. This chapter also studies international and national policies that support sustainable energy in the textile industry. Furthermore, it evaluates the socio-economic advantages like employment, health enhancements and community upliftment in high textile active regions. Despite its potential, the transition has obstacles like high capital costs, technical barriers, and policy gaps. This chapter concludes by recommending integrated renewable energy strategies, supported by regulatory frameworks and technological innovation, to accelerate the textile industry's contribution to climate mitigation.

The sixth chapter is about the renewable energy in the Indian textile sector. The textile industry is one of the most energy intensive sectors globally. A large part of the energy needs in this sector is conventionally sourced through electricity and thermal pipelines. With increasing climate awareness and the need to reduce carbon emissions, the sector is undergoing transformation by adopting renewable energy sources such as solar, wind and hydropower. This chapter explains the energy requirement in the textile manufacturing process for various types of yarn and fibres, and the need to shift to renewable options not just as a measure of green initiative but also to boost the quantity and quality of production significantly. This chapter highlights the textile industry in India—one of the largest textile manufacturers in the world and explores the cost benefit analysis between traditional and renewable energy sources.

I take this opportunity to thank all the contributors for their earnest efforts to bring out this book successfully. I am sure readers of this book will find it very useful.

Renewable Energy and Environmental Sustainability as Lever for Textile Companies' Sustainable Marketing Activities: Insights from a Case Study

Giulia Monteverde and Simone Guercini

Abstract Focusing on the textile sector, this chapter presents the single case study of a medium-sized Italian textile and clothing company. This study highlights how environmental sustainability, proactively implemented by the company primarily through the adoption of renewable energy, can generate what we originally define as a "positive domino effect". The company's sustainability push, being absolutely crucial for counteracting the negative and harmful impacts of the textile industry, initially affects the environmental sustainability domain, which, in turn, creates a positive cascading effect on social and economic sustainability, thus holistically encompassing all sustainable marketing activities. The chapter reveals that the "positive domino effect" is achievable due to the textile company's efforts in environmental sustainability and renewable energy, which have enabled it to strengthen its market reputation while concurrently gaining the trust of numerous involved stakeholders. The company serves as a case in point in this chapter, demonstrating its emergence as a major contributor to sustainability initiatives within the Italian textile landscape, alongside its significant contributions to the local community and the industrial textile district in which it is located. The study offers significant theoretical implications, relevant practical indications for managers and policymakers, and avenues for future research.

Keywords Environmental sustainability · Renewable energy · Textile industry · Case study · Positive domino effect · Sustainable marketing activities · Local community · Industrial district

---

G. Monteverde · S. Guercini (✉)
Department of Economics and Management, University of Florence, Florence, Italy
e-mail: giulia.monteverde@unifi.it; simone.guercini@unifi.it

S. S. Muthu (ed.), *Renewable Energy Implications for the Textile Industry*, Sustainable Textiles: Production, Processing, Manufacturing & Chemistry,
https://doi.org/10.1007/978-3-032-10917-0_2

1 Introduction

In the global economy, the textile industry is recognized as a fundamental sector, along with the apparel sector. In 2023, the global textile market achieved a valuation of USD 1837.27 billion, with projections indicating a compound annual growth rate of 7.4% in revenue during the period 2024–2030; this anticipated market expansion is primarily attributed to the sustained surge in apparel demand within the fashion sector, coupled with the exponential proliferation of e-commerce platforms [1]. Even considering its significant scale, the textile industry is deeply intertwined with the sustainability debate [2, 3]. From a more general overview, the issue of sustainability within the textile and apparel sectors is intrinsically linked to the broader discourse on sustainability in the fashion industry, which increasingly encounters both the opportunities and challenges inherent in sustainable practices [4, 5]. Historically, the textile industry has maintained a profitable trajectory, with its importance significantly augmented by the growth of the fashion industry in recent decades [6]. Specifically, the textile sector demonstrably catalyzes employment generation and economic expansion within low- and middle-income nations. Furthermore, it exerts a critical influence on the elevation of household earnings and the broadening of access to cost-effective apparel for economically marginalized populations [7]. At the same time, the pursuit of sustainable development within the textile industry is characterized by significant complexities; this is particularly true in the case of environmental sustainability.

The textile industry's significant consumption of electricity, fuel, and water results in substantial greenhouse gas emissions and contaminated effluent, posing a serious environmental threat [8]. Textile production processes are associated with considerable energy expenditure and the deployment of chemical agents, thus generating important environmental contamination concerns [9]. This sector also determines notable impact regarding land resource allocation, biodiversity reduction, and the application of high quantities of fertilizers [3]. Energy is one of the most important aspects of the textile industry as it encompasses the whole supply chain, which is heavily energy-intensive and quite long in this sector. With the aim of altering course, a transition to renewable energy is essential for the textile sector to curtail its environmental impact and decouple itself from fossil fuel dependency, paving the way for a greener future. Moreover, as textile companies operate in a business setting of uncertainty and strong competition [10], adopting alternative and novel solutions proves to be determinant, also for the sake of social and economic sustainability.

The textile and apparel industries, constituting a major segment within the fashion sector [11, 12], have significantly contributed to economic and societal advancement in recent decades. This contribution persists despite the industries' confrontation with substantial sustainability challenges [10, 13].

In this scenario, through the analysis of an emblematic single case study [14, 15] in the textile realm, this chapter aims to investigate the potential for textile firms to leverage on the concept of environmental sustainability through the adoption of renewable energy to undertake two directions: (a) firmly establishing themselves in the market as key players in the sustainable transition; and (b) subsequently embracing a holistic approach to sustainable marketing activities. Precisely, sustainable marketing refers to the decision-making process and business activities that undertake all the three dimensions of sustainability, in a triple bottom line perspective [16].

Notably, the triple bottom line framework holds paramount significance, as sustainable companies possess the potential to cultivate broader societal value. This framework can specifically yield enhanced stakeholder value in contrast to the conventional focus on shareholder value. Critically, within the clothing and textile supply chain, the triple bottom line serves as a comprehensive view for developing economic, social, and environmental initiatives while simultaneously fostering sustainable livelihoods. Addressing sustainability imperatives across the entire supply chain, from raw material sourcing to the disposal of finished goods, necessitates a multifaceted strategic response [17].

The case study presented in this chapter focuses on a medium-sized Italian textile and clothing company, founded in the early 1990s. For reasons of confidentiality, the company name is replaced by the first letter of the Greek alphabet, i.e., Alpha. Italy has proved to be a strategic setting to explore the textile sector [18]. Recognizing the urgency of sustainability issues in the contemporary landscape, this company initiated its efforts by targeting the downstream phases of the supply chain. To this end, the company has invested in two new steam generators designed to enhance performance while reducing CO_2 emissions by 5%. Additionally, a cogeneration plant has been installed. Furthermore, Alpha has implemented a patented neutralization system that utilizes CO_2 derived from methane combustion fumes to adjust the pH of wastewater from certain production processes. This textile firm exclusively uses electricity generated from renewable sources, and its production facility is equipped with a rooftop solar panel system. By prioritizing renewable energy, the company has made environmental sustainability one of its distinctive and established features, deeply integrated into the company's values.

This enterprise currently operates on a business model grounded in a circular economy system, relying on renewable energy sources as well as on the selection and transformation of sustainable raw materials, optimization of production processes, and product traceability. The concerted efforts to align with environmental sustainability requirements have allowed Alpha: to strengthen its market reputation, including among fashion industry customers; to become a major contributor to sustainability initiatives within the Italian textile landscape; and to work hard towards achieving legitimacy also in the realm of social and economic sustainable marketing activities, with significant and positive implications for the local area and the industrial district where the company is situated. We advance and originally conceptualize a similar organizational approach as a "*positive domino effect*". The case study analyzed demonstrates how, starting from a pivotal investment in renewable energy as a cornerstone and focusing on the pillar of environmental sustainability, the company was subsequently able to confidently pursue sustainable marketing practices in both the social and economic spheres, becoming a virtuous benchmark for all its stakeholders. Hence, the primary shift towards environmental sustainability implemented by the firm engendered a positive cascade effect, resulting in the comprehensive adoption of sustainability across its three dimensions. Alpha exemplifies a model of virtuous practice for firms in the textile industry aiming to embrace the principles of sustainability.

After the present introduction, the chapter's structure is outlined as follows: Sect. 2 provides the literature review underpinning the study; Sect. 3 elaborates on the methodological approach employed; Sect. 4 presents the case study and key findings, whereas Sect. 5 engages with the research outcomes, and theoretical and managerial implications; ultimately, Sect. 6 concludes the chapter by suggesting avenues for future research.

2 Literature Review

2.1 *Textile Industry's Environmental Sustainability Criticalities*

The textile industry is one of the most resource-intensive sectors globally, contributing significantly to environmental degradation. This impact is primarily driven by the sector's high energy consumption, water usage, pollution generation, and waste production, all of which exacerbate the broader environmental sustainability crisis. The majority of the harmful emissions in this sector is linked to the use of non-renewable energy sources in manufacturing processes such as dyeing, weaving, and finishing, where vast amounts of electricity and heat are required [19]. Particularly, looking into the matter more closely, some pivotal aspects contribute to the environmental sustainability imperative in the textile realm.

One of the most critical and alarming environmental issues is water consumption. The production of textiles requires large amounts of water, a resource already under significant stress in many parts of the world [20, 21]. In addition to water usage, the chemical-heavy dyeing process contributes to water pollution. The environmental burden imposed by the textile industry is attributable to the release of wastewater effluents. These effluents are characterized by a substantial load of organic chemical compounds and colorants, exhibiting limited biodegradability and elevated salinity levels [21]. In this vein, harmful consequences are not only for aquatic ecosystems; indeed, serious health risks to communities dependent on water sources arise.

Moreover, the production of synthetic fibers has added another layer of environmental concern. For instance, polyester, which is derived from petroleum, while being the most widely used fiber in the textile industry, is non-biodegradable, comes from non-renewable resources, and significantly impacts on microplastic pollution. Synthetic textile laundering contributes to the introduction of microfibers into marine environments, leading to accumulation and trophic transfer, which presents a discernible risk to marine life and human health. The protracted environmental persistence of these fibers, extending over decadal timescales, significantly impedes effective mitigation measures and generates long-term environmental vulnerabilities [22]. In addition, other typologies of textile fibers, where wool and cotton can serve as an example, exhibit a high reliance on water and pesticide inputs [23].

The disposal of textiles also contributes to the environmental burden of the industry. Globally, a significant proportion of textiles, surpassing 80%, deviates from the intended closed-loop system upon disposal, resulting in incineration, landfill deposition, or environmental dispersion [24]. The efficacy of current textile collection infrastructures is compromised by fragmentation and capacity constraints, rendering them insufficient for the comprehensive recovery of materials placed within the market [24]. Furthermore, incineration of textile waste produces harmful emissions, further contributing to air pollution and climate change. At the same time, the textile and clothing disposal in landfills generates substantial environmental challenges, primarily due to the non-biodegradable nature of synthetic fibers [25]. In this regard, the growing demand for nylon, polyester, and cotton contributes significantly to the persistent and complex problem of escalating post-consumer textile waste, thereby placing an amplified strain on emerging recycling infrastructure and circular economy approaches [26].

The carbon footprint inherent in the textile industry's transportation and supply chain logistics constitutes another significant environmental liability. The globalized paradigm of textile commerce, characterized by the intercontinental movement of raw materials, manufacturing, and retail distribution, amplifies this impact [27]. In the context of the textile industry's supply chain, and its inherent connection to the fashion sector, the difficulty of establishing traceability is a key concern. In this perspective, the identification of all industrial chemicals utilized and emitted by the textile industry remains a significant challenge, despite widespread recognition of concerns regarding the volume and diversity of chemical inputs. This persistent difficulty mainly stems from capacity limitations, lack of transparency, and inadequate tracking systems [28]. In the case of the fashion industry, traceability emerges as a central concept, involving different actors and encompassing the sharing of complex information about the product and production processes along the entire supply chain [29].

Beyond the environmental issues, the textile industry is further distinguished by deleterious labor practices and exploitative working conditions, which are still inevitably and intrinsically linked to environmental degradation. Although this sustainability dimension predominantly addresses social concerns, it underscores the prioritization of economic gain over both ecological and human well-being in the current textile system [30].

The examination of the environmental sustainability topic in the textile industry necessitates also the inclusion of critical reflections inherent in the fashion sector. In fact, as the demand for fast fashion and mass production continues to rise, all the discussed environmental consequences are expected to worsen. In the current scenario, the fast fashion model and just-in-time manufacturing approach accelerate short-lived trend cycles, driving superfluous consumption and substantial textile waste [31]. In response, the slow fashion movement emphasizes stakeholder needs and tries to minimize environmental and social harm. It promotes sustainability through quality, limited runs, regional sourcing, and equitable labor, moving away from business practices that prioritize the achievement of continuous economic growth. Thus, slow fashion is progressively reshaping textile practices, integrating sustainability across the value chain, often relying on a circular economy model [31–33].

The transition to more sustainable production models, including the use of renewable energy, water-efficient processes, and recyclable materials, is essential for reducing the environmental footprint of the textile industry in the coming decades. Specifically, the following section, in line with the chapter's purpose, mainly focuses on the topic of renewable energy.

2.2 *Addressing Sustainability Urgencies: The Role of Renewable Energy in the Textile Industry*

How can the sustainability criticalities, particularly those of environmental nature, be addressed within the textile industry? Recent academic literature has endeavored to provide an answer to this question, also by contemplating the changes and reactions of the empirical world.

Many companies are embracing the circular economy approach, basing their practices on a regenerative system characterized by the minimization of resource input, waste generation, emissions, and energy leakage through the strategic deceleration, closure, and constriction of material and energy cycles [34]. The transition from linear to circular business models allows to prolong product lifespans and optimize the recycling of used garments [35]. The circular economy system is also particularly linked to the adoption of sustainable innovations, especially in the textile and fashion industries [5, 36, 37]. The pursuit of sustainable product innovation in the textile industry is characterized by a focus on product longevity and performance characteristics, recyclability of materials, and the incorporation of waste into new product design. Process innovation, in parallel, addresses the integration of circular economy models, traceability frameworks, and minimization of water and chemical usage [36]. Sustainable textile innovations can lead to considerable environmental sustainability repercussions [37]. Notwithstanding the articulation of ambitious goals, suppliers of innovative and sustainable textile materials derived from the by-products of different industries, including the food (as in the case of apples) and mining (as in the case of marble) sectors, face specific paradoxes that constrain the comprehensive implementation and diffusion of sustainability principles in the marketplace [5]. However, in light of the undeniable relationship between the textile and fashion sectors, suppliers' support, particularly when they are concentrated within textile industrial districts, is paramount for fashion companies aiming to achieve sustainability outcomes [18, 38]. In this vein, Alpha, the case study presented in this chapter, operates within the Prato textile district, located in the Tuscany region of Italy.

Textile firms are increasingly implementing decarbonization strategies to effectively mitigate carbon emissions across their supply networks [39]. Moreover, the strategic importance of advanced production technologies and information communication technologies in enhancing both production and distribution phases is recognized also within the textile supply chain [40].

In this context, precisely because the textile sector is highly energy-intensive, renewable energy plays a key role. Fossil fuels represent the primary energy source for this industry, reflecting their current market dominance [41]. As a result, the consumption of fossil fuels for the production of electricity and heat within various industrial processes, encompassing textile manufacturing, has experienced a marked increase [42]. Therefore, developing sustainable solutions is crucial and renewable energy represents a viable path for companies. Enhancing circularity and sustainability within the textile industry can be facilitated through the adoption of renewable energy; indeed, alternative clean energy sources present important applications in this sector [43]. Notably, the gradual displacement of fossil fuels and the establishment of renewable energy infrastructure in textile are essential prerequisites [44]. Renewable energy sources present a suitable alternative to fossil fuels, deriving from resources that are inherently inexhaustible and abundant [44]. Considering the textile sector, Hosseini Dehshiri et al. [45] differentiate between electrical energy, primarily utilized for the operations of motors, air conditioning, and lighting, and thermal energy, employed for bleaching, dyeing, and other process applications. Specifically, parabolic solar collectors represent a suitable option for providing thermal energy in the textile industry [46]. Solar thermal energy is an effective form of renewable energy both for electricity and heat. Particularly, a key benefit of solar energy utilization as a clean energy source resides in its capacity to mitigate carbon dioxide emissions [45]. The deployment of solar energy resources in textile manufacturing is characterized by two principal methodologies: solar thermal processing and photovoltaic system implementation. Also, wind energy and biomass energy are other possible alternatives to replace fossil fuels [44].

Hence, the imperative to mitigate anthropogenic carbon dioxide emissions necessitates a transition towards renewable energy sources. Solar, wind, biomass, hydropower, and tidal energy present possible, carbon-neutral alternatives. Substantial reductions in industrial greenhouse gas emissions can be achieved through the integration of renewable energy-based systems. Consequently, a shift from traditional energy supplies to renewable sources, coupled with the development and implementation of novel industrial technologies, is crucial [47]. Among these renewable options, solar power has garnered significant attention due to its abundant, free, and environmentally benign nature, characterized by the absence of noise or pollutant emissions. Extensive research and development (R&D) efforts have focused on enhancing solar energy extraction through technologies such as solar collectors, sun trackers, and concentrated solar power systems, aiming to facilitate its industrial application. As previously mentioned, industrial solar energy applications are broadly categorized into solar thermal and photovoltaic systems. In the context of global energy scarcity and the pressing need to address adverse environmental impacts, the application of solar energy has become a focal point within engineering and economic sciences [47, 48].

The forthcoming 5-year period is projected to witness significant advancements in renewable energy deployment. By 2025, electricity generation derived from renewable sources is anticipated to exceed that of coal-fired power plants. Subsequently, in 2026, both wind and solar energy production are forecasted to individually surpass nuclear power generation. A further pivotal shift is expected in 2029, when solar photovoltaic electricity generation is projected to become the leading global renewable energy source, surpassing hydropower. Concurrently, wind-based generation is anticipated to surpass hydropower by 2030, solidifying the ascendancy of these key renewable energy technologies [49]. Consequently, the textile sector is positioned to be instrumental in the immediate future, realizing the considerable potential inherent in renewable energy integration and environmental footprint mitigation.

2.3 Sustainable Marketing Activities

A comprehensive framework for sustainable marketing activities necessitates the integration of three core dimensions: economic, social, and environmental [50]. Organizational sustainable marketing activities are oriented towards satisfying consumers' socio and ethical needs. These activities demonstrably enhance brand perception, foster consumer trust, and cultivate satisfaction. Consequently, the implementation of sustainable marketing activities determines a favorable brand image and positive consumer attitudes, thereby yielding a competitive advantage for companies [16]. The efficacy of sustainable marketing endeavors is contingent upon the integration of cultural considerations, which are pivotal in fostering societal cooperation and optimizing customer-company communication, ultimately facilitating sustainable organizational growth [51]. Sustainable marketing initiatives also aim to encourage mutual growth and equilibrium across economic, environmental, social, and cultural dimensions [52].

The concept of "sustainable marketing" gives marketing a pivotal role in driving sustainability in the organization. Sustainable marketing is a "*management concept which focuses on achieving the 'triple bottom line' through creating, producing and delivering sustainable solutions with higher net sustainable value whilst continuously satisfying customers and other stakeholders*" ([53], p. 12). Through sustainable marketing, sustainability is incorporated into business strategy across the supply chain, from product and service development to the consumption stage. Hence, marketing departments are tasked with the integration of sustainability principles into product and service offerings, also through the definition of a marketing mix based on sustainability [54]. In this vein, it is crucial that sustainability issues are also shared by employees [54]. For instance, the cultivation of an organizational environmental culture facilitates the development of employee competencies essential for the effective implementation of environmental strategies [55]. Marketing is thus seen as a vehicle to realize the sustainability agenda. The inherent innovative and creative capacities of marketing can serve as a catalyst for the adoption of sustainable practices. Consequently, sustainability must be integrated into the discourse surrounding goods and services consumption, emphasizing economic, social, and environmental value. From an organizational standpoint, a comprehensive and sustainable perspective should be applied to the entire product lifecycle, encompassing design, development, pricing strategies, distribution channels, marketing communications, and end-of-life product and packaging management [54].

Hence, sustainable marketing advocates for a paradigm wherein commercial offerings are promoted responsibly, mitigating adverse impacts on sustainability. This approach prioritizes the marketing of essential goods, addressing fundamental "needs", while de-emphasizing discretionary "wants" [56]. Specifically, Gordon et al. [56] propose a tripartite framework for sustainable marketing, encompassing: (1) green marketing, which focuses on the development and promotion of environmentally sustainable goods and services, embedding sustainability within core marketing and business operations; (2) social marketing, which leverages marketing interventions to foster sustainable individual behaviors across the value chain; and (3) critical marketing, which employs a reflexive analysis of marketing practices to inform regulatory frameworks and stimulate market innovation centered on sustainability, while challenging prevailing systems.

When examining sustainable marketing activities, it is necessary to provide further specification in relation to the three identified components. Firstly, economic (sustainable) marketing activities entail the equitable distribution of economic advantages through regional economic support [16]. These activities focus on value creation and financial performance [57]. A similar corporate approach necessitates the dissemination of economic benefits among consumers and customers, employees, partners, and community stakeholders, while simultaneously fostering organizational growth via profitability. The economic facets of sustainable marketing necessitate the optimization of profitability through managerial efficiency, facilitated by, for instance, the development of e-commerce platforms, the enhancement of retail environments, and the modernization of infrastructural assets [16]. Secondly, the social (sustainable) marketing activities constitute corporate social contribution initiatives that fulfill broader organizational objectives. These activities positively influence consumer perceptions of the company, thereby strengthening customer behavioral intentions, including product evaluations and attitudinal responses. Recognizing their societal role, organizations are expected to discharge their social responsibilities towards the local community [16, 58]. Companies prioritizing the social dimension of their operations engage in marketing initiatives that exert a tangible impact on societal well-being. These activities can encompass philanthropic sponsorships, the cultivation of robust community relations, and the provision of support for educational endeavors [59]. Thirdly, the environmental (sustainable) marketing activities include a spectrum of initiatives aimed at generating and facilitating exchanges that fulfill human needs and desires while minimizing adverse environmental consequences, caring for natural resources [60, 61]. In the textile sector, as investigated in the chapter, these activities can be exemplified by the development of environmentally sustainable materials, produced under an expanding circular economy model, wherein chemical elements are progressively substituted with sustainable components, exemplified by food industry residuals such as apples, pineapples, and oranges [5].

A holistic approach to sustainable marketing necessitates the integration of all three dimensions of the triple bottom line—economic, social, and environmental—rather than reliance on a singular aspect. While the relative emphasis placed on each dimension may vary across industries and marketing contexts, the fundamental importance of these three guiding principles remains consistent [16, 61]. It is also to be acknowledged that the enterprise's size can lead to the implementation of certain sustainable marketing activities rather than others [62]. Thus, companies can rely on sustainable marketing activities as marketing methods to achieve long-term goals and development, while attracting customers and positively impacting on society [63].

In the single case study examined in this chapter, the element of originality highlighted pertains to the fact that a strong corporate valorization and engagement in environmentally focused sustainable marketing activities as yielded a beneficial cascading impact, thereby influencing economic and social sustainable marketing initiatives.

3 Research Methodology and Overview of the Case Study

The chapter employs a qualitative single case study methodology [14, 15] to achieve the research objectives. This exploratory design facilitates an in-depth investigation of context-specific dynamics within a singular organizational setting, allowing for comprehensive inquiry [64, 65]. The application of a qualitative, single case study approach provides a rigorous foundation for scholarly contribution, potentially yielding transferability and comparability across contexts [66]. The company, Alpha, was purposively selected as the focal case due to its significant revelatory capacity, offering unique insights [67]. By employing a case study, the research gains the required depth and breadth to elucidate the specific phenomenon being studied [68].

Alpha is a medium-sized Italian company founded in the early 1990s in Tuscany, specifically within the Prato textile district, which is emblematic of Italian manufacturing excellence and heritage. Alpha is distinguished by its capacity to synthesize traditional craftsmanship with innovative practices, yielding high-quality fabrics and finished garments, while concurrently engaging in collaborations with prominent international fashion brands. The company specializes in the production of premium fabrics and luxury apparel for leading haute couture houses. Alpha exhibits comprehensive operational integration within the textile industry, demonstrating a streamlined management of the entire production and design lifecycle, a feature indicative of its vertical integration. The company's organizational structure encompasses the fabric division, which is focused on fabric research, design, and manufacturing, and the garment division, which is dedicated to garment design, planning, and production, facilitated by an internal sewing laboratory that executes highly technical processes, culminating in the creation of finished garments. It is also noteworthy that the company features its own brand of fashion products. In this perspective, the firm's state-of-the-art production infrastructure, coupled with a highly specialized team, enables the delivery of customized and bespoke solutions, characterized by meticulous attention to quality and detail. Alpha is committed to ensuring the sustainability of its production processes, placing a strong emphasis on technological research and environmental responsibility. The pioneering use of renewable energy sources has consistently been a hallmark of the company, even prior to the current widespread focus on sustainability. As a frontrunner in the adoption of ethical and sustainable production practices, Alpha is dedicated to the continuous reduction of its environmental footprint and the utilization of innovative, eco-friendly materials. Today, Alpha stands as a model of Italian excellence in the sector, demonstrating the capacity to anticipate market trends and respond flexibly to the challenges of the global marketplace. The company has, over time, evidenced its engagement in the creation of a business model that effectively combines innovation and sustainability, with environmental awareness as a core value that has been integral to Alpha's operations from the beginning.

The firm's presentation is inextricably linked to the context in which it operates, namely the Italian textile sector, particularly within the Prato textile district. The Italian textile industry, marked by a dense network of SMEs and a distinguished manufacturing history, operates within a demanding international competitive environment. Notably, this sector is organized around localized industrial districts, such as the Prato district in the case of Alpha, which have a profound economic and social bearing on regional economies [11, 13, 18]. Precisely, Prato, recognized as a historically significant industrial district, has undergone a trajectory of continuous industrial evolution [69, 70]. The district's structure is characterized by a high degree of productive fragmentation among a substantial number of SMEs. Indeed, in response to the fluctuating demands of international markets, textile manufacturers have increasingly outsourced numerous operational processes to specialized local suppliers. The Prato district's industrial ecosystem encompasses several firms specialized in a diverse range of textile products, including yarn for knitting, fabric for apparel, furnishing textiles, nonwoven materials, specialized industrial fabrics, and knitted goods [71]. Since the twelfth century, Prato has maintained a distinguished reputation for its preeminence in textile manufacturing, particularly in woolen fabrics. From the mid-nineteenth century onwards, this industrial district has been recognized for its integrated production system, wherein interconnected, specialized entities contribute to the procurement and processing of raw materials, the advancement of production technologies, and the fabrication of textile goods. In essence, despite the significant transformations that have occurred within the Prato textile district over time, also taking into account entrepreneurial immigration (particularly from China), it is still a significant driving force and a reference point within the Italian fashion industry and landscape of textile entrepreneurship [72].

Considering the data collection phase for the present research, we relied on multiple sources. In doing this, we ensured data triangulation, thereby enhancing the credibility of the data and the construct validity of the research [73, 74]. The core empirical data were derived from face-to-face, in-depth, and semi-structured interviews, serving as the primary source of information [75]. In particular, in May 2024, we conducted one interview with Alpha's CEO and one interview with Alpha's Communication and Marketing Specialist, for a total duration of roughly 2 h. In the same occasion, we had the opportunity to conduct a corporate visit to the Alpha-Hub, which is located in the heart of the Prato industrial district and serves as the management center for the garment division. Specifically, this space also hosts artistic and cultural events, and is made available to the local community for sports and leisure activities. The interviews were audio-recorded and transcribed verbatim. In addition, we collected secondary materials from various sources, such as the company's official website and social media profiles, relevant news publications, videos about the company, sustainability reports, and internal documentation provided by the informants. Moreover, the empirical investigation was facilitated by a pre-existing, longitudinal professional relationship between one of the researchers and the company's management, established through prior research endeavors.

For the data analysis phase, to examine the case study, a content analysis was conducted on both secondary materials and interview transcripts. We carefully analyzed and interpreted the secondary documents and interview transcripts in alignment with the specific research objectives of this chapter [14], trying to focus the attention, firstly, on the corporate commitment to environmental sustainability, underpinned by the utilization of renewable energy and innovative solutions, and, consequently, on the positive repercussions on the company's sustainable marketing activities.

4 Findings from the Case Study

Throughout its operations, Alpha has been able to implement a series of good practices as a fundamental tool to achieve high performance and promote continuous improvement with a view to sustainability, also within the entire textile sector. These practices represent significant experiences and actions that have been shown to achieve better results, both environmentally and socially. In particular, the company has established itself in the market as a key player in the sustainable transition, starting from a deeply rooted environmental commitment, where the adoption of renewable energy has been a cornerstone; subsequently, the commitment has expanded holistically towards all sustainable marketing activities. It is crucial to highlight that Alpha's implemented good practices stem from its unwavering dedication to enhancement and the dissemination of positive internal experiences. Consequently, the company establishes itself as a benchmark for all enterprises engaged in sustainability efforts.

4.1 Environmental Commitment and Renewable Energy Adoption

In its concrete actions towards environmental sustainability, Alpha distinguishes itself through a series of strategies and intervention areas. In particular, the CEO emphasizes that "*environmental sustainability is a fundamental pillar for Alpha. I recall that, 10 to 15 years ago, I had the desire to purchase a dry-cleaning machine that utilized perchloroethylene, one of the most harmful substances known. Although these machines operate on a closed-loop system, leakage is inevitable, and pollution is generated as perchloroethylene is, in effect, a potent toxin. At that time, I opted for a hydrocarbon-based machine, which, while possessing a lower cleaning efficacy, was environmentally sustainable. When I acquired it, I had to source it from Germany, as no Italian manufacturer could provide quality assurances for this type of machinery. This approach has always been natural and spontaneous for me ...*".

Specifically, particular attention is given to energy-related issues. In this regard, the company is committed to utilizing renewable energy sources, such as solar power, to fuel its operations. Particularly, to generate clean energy from solar resources, the company installed photovoltaic arrays on the production facility's rooftop. Additionally, an agreement was secured with an energy provider, ensuring that all electricity supplied to the company is derived exclusively from renewable sources, including hydroelectric, wind, and solar power. Furthermore, in this perspective, energy-saving measures are implemented, including the installation of low-energy lighting systems and the optimization of production processes to minimize energy waste. Alpha has made significant investments in new high-performance steam generators, enhancing the overall efficiency of the production process and reducing CO_2 emissions by 5%. Also, it has developed and patented a neutralization plant that utilizes CO_2 derived from methane combustion flue gases. This plant plays a crucial role in treating wastewater from certain processing stages. The CO_2 is employed to correct the pH of the wastewater, rendering the treatment process more efficient and sustainable. Investments in new steam generators and the neutralization plant enable the company to improve its environmental performance and contribute to the fight against climate change. In this respect, to summarize some preceding lines of reasoning, the CEO points out as follows: "*the company undertakes numerous initiatives aimed at energy efficiency optimization. For instance, all premises have undergone renovation, and all window and door frames feature thermal break technology for enhanced insulation. The heating systems are all underfloor, operating with radiant heat, suitable for both summer and winter. The entire design is, shall we say, meticulously planned ... We have extensive solar panel installations and have exclusively utilized recycled paper for many years, practically since inception*".

Responsible water resource management is another crucial element of Alpha's planet-friendly actions. The firm develops practices to reduce water consumption in its operations, for example through wastewater recycling and recovery systems. Moreover, it promotes the sustainable use of water resources in the local communities where it operates, collaborating with organizations and institutions for water resource conservation and water pollution prevention. CO_2 emission reduction is a fundamental priority for Alpha. This company in fact implements greenhouse gas emission reduction strategies, adopting advanced technologies and low environmental impact production processes. Moreover, it promotes the use of eco-friendly transportation for its logistics operations, thus contributing to the overall reduction of transport-related emissions.

Supporting environmental awareness initiatives is an additional aspect of the firm's environmental actions. Indeed, the company collaborates with environmental organizations and actively participates in projects aimed at environmental protection and biodiversity conservation. These efforts include, for instance, participation in environmental cleanup campaigns, reforestation projects, and initiatives for the protection of local ecosystems.

Considering the production process, as the CEO underlines, "*the utilization of natural dyes enables a sustainable production process*", facilitated by the company's internal research and development (R&D) department which has successfully industrialized the process on vegetable fibers.

Of significant interest is the implementation, in the early 2020s, of a textile waste repurposing system, facilitating the creation of novel apparel within a circular production paradigm. The re-integration of fibers into the manufacturing cycle has a substantial impact, not only on manufacturing policy, but especially on improving the global environmental quality. This Alpha project allows its customers and partners to participate in the collection and recovery of all textile waste. Notably, the recycling of textile waste represents a considerable contribution to the creation of a culture of environmental responsibility to be shared with the companies Alpha collaborates with. The project focuses on the creation of new products through the recovery of textile materials, which may be fabrics stored in warehouses or finished products, which are then divided into unsold, returned, or end of life products. One of the distinguishing features of the resulting products is their certified traceability, from the collection and logistics phase to delivery. This initiative implemented by the enterprise allows for the complete elimination of textile waste disposal costs, estimated at approximately 250–300 € per ton.

Therefore, in today's landscape, environmental consciousness has invariably been a key focus for Alpha, which has firmly adopted this concern, evolving into a reference point within its field by deploying environmental schemes intended to lessen its carbon footprint. Particularly, it is paramount to further emphasize that Alpha's dual approach to renewable energy represents an exemplary model for the transition towards sustainability within the textile sector. The approach can be termed "dual" as the company strategically combines on-site energy generation, through the installation of photovoltaic panels on the factory roof, with the procurement of green energy from a certified supplier. This synergy enables Alpha to maximize the utilization of renewable sources, significantly reducing its reliance on fossil fuels and, consequently, its carbon footprint. The analysis of energy consumption and CO_2 emissions data reveals a clear positive correlation between the implementation of these practices and the diminution of the firm's environmental impact.

Specifically, a measurable reduction in greenhouse gas emissions is observed, paralleling the increase in the proportion of renewable energy utilized. This integrated approach not only demonstrates Alpha's commitment to sustainability but also provides a replicable framework for other companies within the textile industry that aim to lessen their environmental impact.

4.2 Sustainability in the Company's Supply Chain

Sustainable considerations within the supply chain are crucial for Alpha, which maintains a close partnership with its suppliers to enforce strict ethical and environmental standards. This dedication is reflected in the meticulous selection of suppliers, emphasizing sustainability, regulatory compliance, and respect for human rights. Alpha also conducts periodic audits of supplier facilities to monitor labor conditions, ensuring the protection of workers' rights and workplace safety. Considering the words of the Communication and Marketing Specialist during the in-depth interview, "*our supply chain is characterized by controlled, traceable, and transparent operations. […] We place significant emphasis on communicating this externally*".

Considering a more comprehensive view, Alpha meticulously manages relationships within its productive supply network. In addition to suppliers, careful selection is applied to all trade partners, as well as entities and firms managing components of the productive processes via outsourcing. Specifically, the firm chooses partners based on professionalism, the quality of provided products and services, and the prompt management of requests. Despite internalizing the majority of operations, certain key processes necessitating specialized expertise are delegated to external suppliers, typically local entities, including those within the same Prato textile district.

The selection of collaborators with whom to share ideals and values of sustainability also concerns company employees, to create a shared corporate culture based on the assumption of responsibility, starting with environmental responsibility. Cultivating a pervasive culture of sustainability, such as that centered on renewable energy, requires the enterprise's implementation of comprehensive employee engagement programs that promote awareness and participation. Additionally, the company advocates for consumer transparency by providing detailed information on the supply chain and material provenance. Hence, "*we, as an enterprise, perceive ourselves as genuinely responsible for our supply chain and diligently monitor the operations of all stakeholders involved. We endeavor to communicate that the creation of habitable, clean, and organized work environments fosters employee enthusiasm and dedication. Consequently, employees work with passion, meticulousness, and precision, cultivating a sense of belonging to the company. This involves the transmission of a culture of responsibility, which is essential for the advancement of the entire supply chain*", as the CEO states; in this vein, the informant continues: "*if I am responsible for the sewing workshop that works for my company, then I have to insist that it meets certain requirements. By doing this, I am not just helping the workshop itself, but I am also helping its employees, who then in turn help the workshop, which ultimately benefits my company. It is a cycle of positivity that spreads*".

The commitment to supply chain ethics, and Alpha's role in establishing the legitimacy of good practices throughout the supply chain, is a direct result of robust credibility the company has cultivated and maintained. With reference to this, Alpha has now garnered the trust and appreciation of its luxury fashion clientele as well, by offering innovative and sustainable solutions that consistently consider Italian craftsmanship and the tradition of the Prato textile district.

4.3 Societal and Economic Engagement Through the Company's Contribution to the Local Community

Alpha is deeply rooted in its native territory, the Prato textile district. This territorial link extends to the local community, encompassing its people, institutions, various stakeholders, and environmental surroundings. Over a period of time, through determined, structured, and successful undertakings in environmental sustainability, the company has become a reference point, initially within the local context, and subsequently at the regional and national scales. The enterprise's initial focus on environmental sustainability has generated a robust foundation, enabling a more holistic intervention into social and economic sustainability activities.

Specifically, Alpha has established a trustworthy relationship with the local community, which recognizes its authentic role in ensuring the economic sustainability of the entire region. The CEO illustratively states: "*at Alpha, we have a significant number of young employees, both male and female; in fact, women make up roughly 47% of our workforce ... This is not typical for a textile company! Young people bring a lot of enthusiasm and energy. Above all, when they grow within a company, they develop a strong sense of ownership and embrace the company culture wholeheartedly*". The mutual dependence between the company and the local ecosystem presents a potential for substantial synergies. Alpha is currently backed by various entities that have been brought together through corporate activities and recreational initiatives, which have served as a cohesive force in establishing solid bonds. The company promotes numerous events, demonstrations, donations, and ethical initiatives to solidify its value system. Additionally, the company's objectives include the redistribution of generated value to the local community, and the development of progressively innovative processes aimed at reducing environmental impact. The work environment at Alpha is appealing as it is both dynamic and modern. In the interview, the Communication and Marketing Specialist explained that the company's social and economic sustainability efforts are a key feature, further supported by an already solid corporate reputation in environmental sustainability: "*in addition to events, numerous projects we organize pursue educational goals. Our social sustainability initiatives, inherently spontaneous, are a defining characteristic, solidified over time and further enhanced by the company's favorable reputation*".

The center of social and environmental actions is the Alpha-Hub, a design center—inaugurated in the early 2020s—that symbolizes the company's commitment to the environment and people, as well as the constant effort to improve company processes. Alpha's activities include the training and professional development of staff, special attention to issues of diversity and inclusion, and the provision of services to the community. The Alpha-Hub serves as a design space for quality Made in Italy products and a space for the company's engagement with the city. In this sense, the CEO' words are truly emblematic: "*at Alpha-Hub, we consistently organize diverse initiatives, we have a sports field within the company where it is possible to engage in sports; we switch from one sport to another! Last year we played padel, this year we are playing basketball. The company is open to everyone. [...] We frequently organize, like a few evenings ago, wheelchair basketball games. [...] The neighborhood around Alpha-Hub is flourishing since we arrived. The company is integrated with the Prato community, so it is important to stimulate the community and the community must know that there is a company it can rely on*".

The Alpha-Hub serves as a venue for the convergence and preservation of the local district's cultural and productive heritage. It offers a novel work model, integrating production, training, sales, and space utilization within a unified framework, aiming to foster synergy among diverse activities, stimulate collaboration among textile enterprises, and promote the growth of the textile sector in Prato. Consequently, the Alpha-Hub positions itself as a focal point not only for industry professionals but also for the general public. Designed as an "*open space, replete with glass windows and natural light*", as emphasized by the CEO during the interview, it allows individuals to undertake tours to observe the artisanal processing of fabrics and produced garments. The Alpha-Hub is accessible to citizens, training institutions, customers, designers, and all textile industry operators, creating opportunities for work and growth. The space can be utilized to organize cultural, artistic, and sporting events, including hosting prominent figures, as it has been done in previous years. Furthermore, the Alpha-Hub is a reference point for organizing workshops, training courses, and managing collective projects with local associations.

In essence, the Alpha-Hub aspires to become even more a pivotal reference point for the local community by offering an innovative and flexible service while concurrently committing to contributing to societal well-being and economic sustenance. This objective is achievable solely through collaborative engagement with the community and the activities undertaken within its multi-purpose space, wherein the active participation of citizens represents the linchpin for the success of this long-term social project.

5 Discussion and Theoretical Implications: Advancing a "Positive Domino Effect"

The findings of our study allow for the derivation of several significant conclusions, concurrently contributing to the theoretical debate and yielding practical and managerial implications.

The textile company's effective alignment with environmental sustainability imperatives has demonstrably enhanced its market reputation, particularly among customers within the fashion sector. This proactive engagement has positioned the firm as a salient contributor to sustainability initiatives within the Italian textile industry. Furthermore, the company is actively pursuing legitimacy in the domain of social and economic sustainable marketing activities, generating substantial positive ramifications for the local community and the broader industrial cluster in which it operates. Indeed, this case study illustrates how a foundational investment in renewable energy, serving as a central tenet of environmental sustainability, has enabled the company to subsequently and confidently implement sustainable marketing practices across social and economic dimensions comprehensively, thereby establishing a virtuous benchmark for its diverse stakeholders.

Based on this chapter's findings, it is possible to advance that the initial and demonstrable commitment of firms to environmental sustainability, particularly through the substantive integration of renewable energy sources into their operational frameworks, can serve as a critical catalyst for a cascading series of positive impacts. This foundational environmental responsibility not only directly mitigates ecological footprints but also cultivates a robust basis for enhanced corporate legitimacy and reputational capital. As stakeholders increasingly prioritize environmentally conscious practices, companies proactively embracing renewable energy gain a significant advantage in building trust and credibility. This enhanced reputation, initially anchored in environmental responsibility, subsequently paves the way for successful engagement in broader sustainability domains, encompassing social and economic viability. Consequently, the adoption of renewable energy can act as a pivotal first step, enabling companies to authentically embrace comprehensive sustainable marketing activities and foster long-term value creation across the three sustainability dimensions. Our reasoning is presented as a "positive domino effect", as visually showed in Fig. 1, where embracing sustainable marketing activities starts from environmental sustainability and renewable energy implementation.

Within the textile industry, a sector historically associated with significant environmental impacts [19, 26], the proactive and deeply embedded commitment of a company to environmental sustainability, with a central tenet being the substantial integration of renewable energy sources [47], represents a strategic imperative. This fundamental dedication to environmental responsibility yields direct and measurable reductions in the textile industry's considerable ecological footprint, encompassing decreased greenhouse gas emissions from energy-intensive processes like dyeing and finishing, diminished reliance on fossil fuels for power generation in production facilities, and a lessened burden on water resources and chemical usage

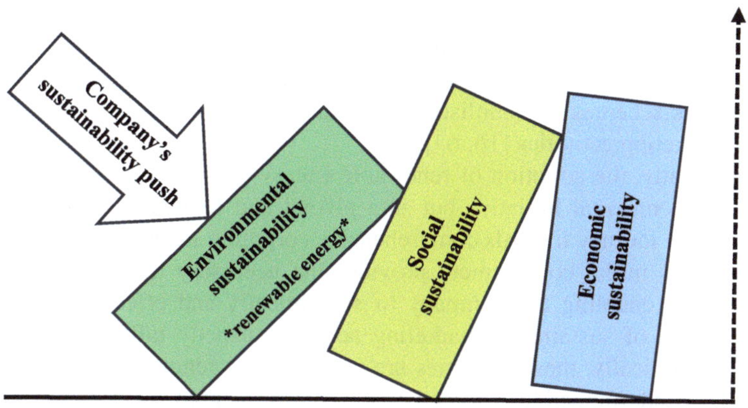

Fig. 1 Conceptualization of a "positive domino effect". (Source: Authors' elaboration)

often linked to traditional textile manufacturing. However, its significance transcends these tangible environmental benefits, concurrently establishing a robust framework for the accretion of enhanced corporate legitimacy and the cultivation of reputational capital within an increasingly discerning and environmentally aware global fashion and textile market [4, 5].

Since a diverse array of stakeholders progressively prioritize and actively demand demonstrably environmentally conscious practices, textile companies that proactively and authentically embrace renewable energy technologies and integrate them as core components of their operational paradigms accrue a significant and often decisive competitive advantage in the arena of building and sustaining stakeholder trust and credibility. In particular, these stakeholders can include fashion customers looking for sustainable textile suppliers, conscious consumers demanding eco-friendly clothing, investors increasingly scrutinizing the environmental performance of fashion brands, stringent regulatory bodies imposing stricter environmental standards on the textile sector, ethically minded employees seeking alignment with sustainable values, and communities impacted by textile production. As depicted in Fig. 1, the "positive domino effect" underscores the role of environmental sustainability as a significant and essential catalyst, initiating a comprehensive and iteratively strengthening sequence of positive consequences that transcend the immediate ecological benefits associated with renewable energy adoption.

The enhanced corporate reputation and elevated level of stakeholder trust anchored in environmental responsibility have the potential to allow companies to operate in broader and more encompassing sustainability domains relevant to the fashion and textile industry. This positive reputational spillover effect allows the company to more effectively address and integrate the interconnected pillars of social and economic responsibility into its overarching sustainability framework. For instance, following the case study presented in the chapter, a textile company recognized for its environmental commitment through renewable energy adoption is often better positioned to advocate for fair labor practices within its complex supply chain, ensure safe and healthy working conditions in its manufacturing facilities, promote the use of sustainably sourced and ethically produced raw materials, and develop innovative circular economy initiatives such as textile recycling programs and take-back schemes. This holistic approach comprehensively encompasses sustainable marketing activities [16, 63].

Consequently, the adoption of renewable energy sources acts not merely as an isolated environmental initiative but as a pivotal and foundational first step in a transformative journey towards comprehensive corporate sustainability for a textile company. This initial commitment provides the necessary legitimacy and stakeholder buy-in, enabling the company to authentically and effectively embrace a wide spectrum of sustainable marketing activities exactly tailored to the textile industry. Specifically, these activities are perceived as genuine and credible since they are also underpinned by a demonstrable commitment to environmental responsibility. A similar integrated approach to sustainability fosters long-term value creation across the triple bottom line, ultimately contributing to a more sustainable future for the clothing and textile industry, and its various stakeholders. Indeed, the positive feedback loops generated by the initial environmental commitment and subsequent broader sustainability engagement create a fertile ground for a cycle of continuous improvement for the textile sector in a rapidly evolving market.

In addition, this research highlights that the broader textile district, as the Prato industrial district (in the specific case study analyzed), and the relevant local community benefit from the "positive domino effect" generated by the company's sustainability push (see Fig. 1). First of all, the textile company can serve as a positive benchmark and inspire other companies within the textile district to adopt similar sustainable marketing activities. It showcases that environmental responsibility is not only feasible but also strategically advantageous within the sector. Moreover, when a company within the textile district gains recognition for its sustainability efforts, it positively reflects on the entire district. Thus, this can enhance the district's overall reputation, possibly making it more attractive to local, national, and international buyers, investors, and talent. In this vein, a district positively known for its commitment to sustainability can also attract other sustainability-focused businesses, fostering collaborations and partnerships aimed at creating a more sustainable supply chain; relationships could also be established with non-business actors such as associations and educational institutions [38]. The adoption of renewable energy and the subsequent pursuit of broader sustainability initiatives often necessitate innovation in processes, technologies, and materials. The success of one company can lead to the sharing of knowledge and best practices within the entire district, fostering an approach of continuous improvement and competitiveness, and supporting sustainable innovations [76]. On the other side, considering the local community, the direct environmental benefits of a company using renewable energy, such as reduced air and water pollution and a smaller carbon footprint, directly contribute to a healthier and more sustainable local environment for the community. In addition, the firm's strengthened reputation can lead to greater economic stability and potential job creation within the community. Moreover, the focus on sustainability can spur the development of new green jobs (e.g., sustainability manager, textile waste recycling specialist, energy manager, and environmental consultant) and skills within the local workforce. From a general point of view, a company committed to environmental, social, and economic sustainability is more likely to be a responsible corporate citizen, contributing positively to the overall quality of life in the local community [77]. When a local company in the district is recognized for its sustainability efforts, it can generate a sense of pride and positive identity within the community. Finally, the company's sustainability initiatives can create advantageous opportunities for collaborations with local educational institutions, research centers, and community organizations on environmental projects and sustainability awareness programs.

5.1 *Implications for Companies and Policymakers*

Focusing on more practical implications, we state that the "positive domino effect" initiated by a textile company's commitment to renewable energy has significant implications for how companies are managed and how policies are formulated. Managers need to embrace a strategic, long-term perspective that integrates environmental sustainability into core business operations, while policymakers have a crucial role in creating an enabling environment that incentivizes and supports the widespread adoption of renewable energy and other sustainable practices within the textile sector, ultimately benefiting both the industry and the wider community.

Managers should recognize that investing in renewable energy is not merely an operational cost but a strategic investment that yields multifaceted returns: it forms a building block and this necessitates a shift from viewing environmental initiatives as compliance burdens to recognizing them as core value drivers. Also, companies need to transparently communicate their commitment to renewable energy and the resulting environmental benefits to all stakeholders. This includes clear reporting on energy sources, emissions reductions, and other sustainability metrics. Indeed, as previously emphasized, engaging with stakeholders builds trust and reinforces the company's legitimacy. Moreover, as embracing renewable energy often necessitates exploring and adopting new technologies and processes, managers should foster a culture of innovation within their organizations to identify and implement sustainable solutions across the value chain, from energy sourcing to material selection and waste management. Accordingly, managers should actively collaborate with suppliers and other business players to promote a collective sustainability responsibility. It is furthermore crucial that companies recognize the extent to which environmental sustainability initiatives can positively impact the entirety of their sustainable marketing activities.

Conversely, within this perspective of transition towards sustainability, policymakers also occupy a pivotal role. They should implement policies that incentivize the adoption of renewable energy by textile companies. This can include tax breaks, subsidies, grants, and streamlined permitting processes for renewable energy installations. Clear and consistent environmental regulations, including targets for emissions reduction and renewable energy use, can provide a level playing field and drive the adoption of sustainable practices across the textile sector. Additionally, policies should encourage collaboration and knowledge sharing among companies in the industrial district on environmental best practices. Also, public funding should be allocated to support research and development (R&D) in sustainable textile technologies, including renewable energy applications in textile manufacturing, innovative eco-friendly materials, and circular economy solutions. It is advisable for policymakers to implement clear, direct, and decisive interventions. Public investment in infrastructure that supports renewable energy deployment, such as grid modernization and the development of renewable energy sources within the region, can facilitate the transition to cleaner energy for textile companies and other industries. Policymakers should invest in training programs that equip the workforce with the skills needed for the green transition in the textile sector, including the installation, maintenance, and operation of renewable energy systems and sustainable manufacturing technologies. Coordinating dialogue and collaboration is another critical area as policymakers should organize ad hoc occasions for meetings between textile companies, research institutions, NGOs, and community stakeholders with the objective of identifying challenges and potential opportunities for advancing sustainability within and beyond the district.

6 Future Research Prospects

The textile sector has yielded particularly insightful observations regarding the interconnectedness of sustainability dimensions and sustainable marketing activities. However, future research would benefit from extending this line of inquiry to other industries to ascertain the generalizability of the findings presented in this chapter. Specifically, it would be valuable to investigate whether the observed positive cascading effect, initiated by a primary focus on environmental sustainability, also manifests in other sectoral contexts and, if so, to delineate the modalities and characteristics of this diffusion. This would help identify sector-specific nuances, the types of initial sustainability drivers that are most effective in different contexts, and the unique pathways through which the positive spillover unfolds.

A potentially fruitful avenue of exploration would be to examine whether the "positive domino effect" can be triggered initially by prioritizing one of the other dimensions of sustainability, such as social or economic sustainability, and to subsequently analyze the resultant impacts on the remaining facets of the sustainability triad. Investigating such alternatives could provide a more nuanced understanding of the interrelationships between environmental, social, and economic sustainability across diverse industries.

While this book chapter's analysis is grounded in a single, albeit emblematic, case study that effectively illuminates the phenomenon under investigation, the inherent limitations of a singular instance necessitate future scholarly endeavors to adopt a more expansive methodological approach. Subsequent research should prioritize the development of a cross-case comparative analysis. Hence, future studies can enrich the theoretical framework presented herein, focusing on the understanding of the dynamics at play.

Future research could also employ longitudinal designs to track the evolution of the "positive domino effect" over extended periods. This would allow for a deeper understanding of the temporal dynamics, potential time lags between initial environmental actions and subsequent social and economic benefits, and the long-term sustainability of these effects.

Moreover, it could be valuable to thoroughly investigate the moderating role of stakeholder engagement in facilitating or hindering the "positive domino effect". For instance, the corporate level and quality of engagement with customers, investors, employees, and the local community can influence, maybe with diverse intensity, the translation of environmental sustainability into social and economic benefits. At the same time, the internal factors within a company, such as its culture and the leadership's commitment to sustainability, can play a significant role in driving and amplifying the "positive domino effect". We need to consider that subsequent scholarly investigations could be enriched by including also an analysis of external factors, such as government regulations, market pressures, and societal norms; indeed, they can impact on the emergence and strength of the "positive domino effect". Comparative studies across different policy environments could provide important insights for policymakers.

Before concluding this chapter, one key point warrants consideration. An important avenue for future research lies in investigating possible cases where the anticipated positive spillover effects from an initial focus on one sustainability dimension fail to materialize or, more concerningly, result in negative repercussions across other areas. While the present study underlines the potential for a "positive domino effect", an examination of the conditions under which this effect happens (or not) would be valuable. Future research should therefore seek to identify the contextual, organizational, or industry-specific factors that may impede the transmission of positive outcomes from, for example, environmental initiatives to social or economic spheres. Furthermore, exploring scenarios where a singular focus on one aspect of sustainability inadvertently creates trade-offs or negative externalities in others, such as environmentally focused initiatives that negatively impact local employment or socially driven programs that strain economic viability, is critical. Exploring the potential for unintended negative consequences can be fundamental for developing more effective sustainable marketing activities.

In any case, the integration of renewable energy sources into the textile industry not only presents a feasible pathway towards mitigating its significant environmental impact and fostering a future characterized by enhanced sustainability, resource independence, and the potential for circular economic models, but also generates a catalytic domino effect, positively influencing the scope of companies' sustainable marketing activities across the entire value chain.

Acknowledgements This research was supported by Piano Nazionale di Ripresa e Resilienza, Missione 4—Componente C2—Investimento 1.1 Fondo per il Programma Nazionale di Ricerca e

Progetti di Rilevante Interesse Nazionale (PRIN), finanziato dall'Unione Europea—NextGenerationEU—Bando PRIN 2022 DD MUR n. 104 del 02-02-2022—Progetto "Communicating Companies' Supply Chain Sustainability Practices in a Digital Environment" n. progetto PRIN 2022LHHLWM—CUP B53D23010060006.

References

1. Grand View Research (2024) Textile market size, share & trends analysis report. https://www.grandviewresearch.com/industry-analysis/textile-market. Accessed 17 April 2025
2. Muthu SS, Gardetti MA (2020) Sustainability in the textile and apparel industries. Springer, Cham
3. Xu W, Jia FJ, Chen L, Schoenherr T (2024) Sustainable transition in textile and apparel industry. J Clean Prod 443:141081
4. Akrout H, Guercini S (2022) Sustainability in fashion and luxury marketing: results, paradoxes and potentialities. J Glob Fash Market 13(2):91–100
5. Runfola A, Monteverde G, Guercini S (2025) Sustainable innovations in business-to-business at the crossroad: emerging paradoxes in the fashion industry. J Bus Ind Mark 40(13):87–102
6. Abbas S, Hsieh LHC, Techato K, Taweekun J (2020) Sustainable production using a resource–energy–water nexus for the Pakistani textile industry. J Clean Prod 271:122633
7. Eppinger E (2022) Recycling technologies for enabling sustainability transitions of the fashion industry: status quo and avenues for increasing post-consumer waste recycling. Sustain Sci Pract Policy 18(1):114–128
8. Hasanbeigi A, Price L (2015) A technical review of emerging technologies for energy and water efficiency and pollution reduction in the textile industry. J Clean Prod 95:30–44
9. Reike D, Hekkert MP, Negro SO (2023) Understanding circular economy transitions: the case of circular textiles. Bus Strategy Environ 32(3):1032–1058
10. Cavalcanti Sá de Abreu M, Ferreira FNH, Proenca JF, Ceglia D (2021) Collaboration in achieving sustainable solutions in the textile industry. J Bus Ind Mark 36(9):1614–1626
11. Guercini S, Runfola A (2012) Relational paths in business network dynamics: evidence from the fashion industry. Ind Mark Manag 41(5):807–815
12. Runfola A, Milanesi M, Guercini S (2022) Internationalization of luxury fashion firms, Examining the business models of SMEs. Palgrave Macmillan, Cham
13. Ferlito R (2024) Industry 4.0 and sustainability: the case of the Italian textile district of Prato. Compet Rev 34(5):995–1016
14. Eisenhardt KM (1989) Building theories from case study research. Acad Manage Rev 14(4):532–550
15. Yin RK (2009) Case study research: design and methods. Sage, Thousand Oaks
16. Jung J, Kim SJ, Kim KH (2020) Sustainable marketing activities of traditional fashion market and brand loyalty. J Bus Res 120:294–301
17. Hiller Connell KY, Kozar JM (2017) Introduction to special issue on sustainability and the triple bottom line within the global clothing and textiles industry. Fashion Text 4:1–3
18. Guercini S (2004) International competitive change and strategic behaviour of Italian textile-apparel firms. J Fash Mark Manag 8(3):320–339
19. Desore A, Narula SA (2018) An overview on corporate response towards sustainability issues in textile industry. Environ Dev Sustain 20:1439–1459
20. Alkaya E, Demirer GN (2014) Sustainable textile production: a case study from a woven fabric manufacturing mill in Turkey. J Clean Prod 65:595–603
21. Vajnhandl S, Valh JV (2014) The status of water reuse in European textile sector. J Environ Manage 141:29–35

22. Deng H, Wei R, Luo W, Hu L, Li B, Di Y'N, Shi H (2020) Microplastic pollution in water and sediment in a textile industrial area. Environ Pollut 258:113658
23. Caniato F, Caridi M, Crippa L, Moretto A (2012) Environmental sustainability in fashion supply chains: an exploratory case based research. Int J Prod Econ 135(2):659–670
24. Ellen MacArthur Foundation (2024) We need extended producer responsibility (EPR) policy for textiles. https://www.ellenmacarthurfoundation.org/epr-policy-for-textiles
25. Fedorak P (2006) Microbial processes in the degradation of fibers. In: Blackburn RS (ed) Biodegradable and sustainable fibres. Woodhead, pp 1–35
26. Dziubaniuk O, Aarikka-Stenroos L (2025) Ethical value co-creation in circular economy ecosystems: a case study of the textile industry. J Bus Ind Mark 40:1423–1438
27. Borghesi S, Vercelli A (2003) Sustainable globalisation. Ecol Econ 44(1):77–89
28. United Nations Environment Programme (UNEP) (2023) Sustainability and circularity in the textile value chain—a global roadmap. UNEP, Paris
29. Guercini S, Runfola A (2009) The integration between marketing and purchasing in the traceability process. Ind Mark Manag 38(8):883–891
30. Annapoorani SG (2017) Social sustainability in textile industry. In: Muthu SS (ed) Sustainability in the textile industry. Springer, Singapore, pp 57–78
31. Pookulangara S, Shephard A (2013) Slow fashion movement: understanding consumer perceptions—an exploratory study. J Retail Consum Serv 20(2):200–206
32. Fletcher K (2010) Slow fashion: an invitation for systems change. Fash Pract 2(2):259–265
33. Jia F, Yin S, Chen L, Chen X (2020) The circular economy in the textile and apparel industry: a systematic literature review. J Clean Prod 259:120728
34. Geissdoerfer M, Savaget P, Bocken NM, Hultink EJ (2017) The circular economy–a new sustainability paradigm? J Clean Prod 143:757–768
35. Hartley K, Roosendaal J, Kirchherr J (2022) Barriers to the circular economy: the case of the Dutch technical and interior textiles industries. J Ind Ecol 26(2):477–490
36. Dominidiato M, Guercini S, Milanesi M, Tunisini A (2023) Supplier-customer relationships for sustainability-led innovation in the textile industry. J Bus Ind Mark 39(13):15–26
37. Rese A, Baier D, Rausch TM (2022) Success factors in sustainable textile product innovation: an empirical investigation. J Clean Prod 331:129829
38. Runfola A, Monteverde G (2023) Something new, something sustainable! How network relationships shape sustainable new venture development. J Bus Ind Mark 38(12):2811–2830
39. Guo H, Dong M, Tsinopoulos C, Xu M (2024) The influential capacity of carbon neutrality environmental orientation in modulating stakeholder engagement toward green manufacturing. Corp Soc Respon Environ Manag 31(1):292–310
40. Karmaker CL, Al Aziz R, Ahmed T, Misbauddin SM, Moktadir MA (2023) Impact of industry 4.0 technologies on sustainable supply chain performance: the mediating role of green supply chain management practices and circular economy. J Clean Prod 419:138249
41. Hurmekoski E, Kunttu J, Heinonen T, Pukkala T, Peltola H (2023) Does expanding wood use in construction and textile markets contribute to climate change mitigation? Renew Sustain Energy Rev 174:113152
42. Khandaker S, Bashar MM, Islam A, Hossain MT, Teo SH, Awual MR (2022) Sustainable energy generation from textile biowaste and its challenges: a comprehensive review. Renew Sustain Energy Rev 157:112051
43. Kumar KR, Chaitanya NK, Kumar NS (2021) Solar thermal energy technologies and its applications for process heating and power generation–a review. J Clean Prod 282:125296
44. Farhana K, Kadirgama K, Mahamude ASF, Mica MT (2022) Energy consumption, environmental impact, and implementation of renewable energy resources in global textile industries: an overview towards circularity and sustainability. Mater Circ Econ 4(15):1–16
45. Hosseini Dehshiri SS, Hosseini Dehshiri SJ, Firoozabadi B (2023) Evaluation of using solar energy in Iran's textile industry towards cleaner production: sustainable planning and feasibility analysis. J Clean Prod 421:138447

46. Carnevale EA, Ferrari L, Paganelli S (2011) Investigation on the feasibility of integration of high temperature solar energy in a textile factory. Renew Energy 36(12):3517–3529
47. Mekhilef S, Saidur R, Safari A (2011) A review on solar energy use in industries. Renew Sustain Energy Rev 15(4):1777–1790
48. Bazen EF, Brown MA (2009) Feasibility of solar technology (photovoltaic) adoption: a case study on Tennessee's poultry industry. Renew Energy 34(3):748–754
49. International Energy Agency (2024) Renewables 2024: analysis and forecasts to 2030. IEA, Paris. https://www.iea.org/reports/renewables-2024. Accessed 17 April 2025
50. Elkington J (1998) Cannibals with forks: the triple bottom line of 21st century business. New Society, Stony Creek
51. Ko E, Chae H, Chun E, Yong S, Kong HM, Han J (2015) Sustainable fashion brand marketing. Kyomunsa, Paju
52. Gladwin TN, Kennelly JJ, Krause TS (1995) Shifting paradigms for sustainable development: implications for management theory and research. Acad Manage Rev 20(4):874–907
53. Charter M, Peattie K, Ottman J, Polonsky MJ (2002) Marketing and sustainability. Centre for Business Relationships, Accountability, Sustainability and Society (BRASS) in Association with The Centre for Sustainable Design, pp 1–36
54. Jones P, Clarke-Hill C, Comfort D, Hillier D (2008) Marketing and sustainability. Mark Intell Plan 26(2):123–130
55. McDonagh P, Prothero A (2014) Sustainability marketing research: past, present and future. J Mark Manag 30(11/12):1186–1219
56. Gordon R, Carrigan M, Hastings G (2011) A framework for sustainable marketing. Mark Theory 11(2):143–163
57. Bansal P (2005) Evolving sustainably: a longitudinal study of corporate sustainable development. Strateg Manag J 26(3):197–218
58. Lee MY, Sung J (2016) Sustainability and management in fashion, design and culture. J Glob Fash Market 7(2):73–75
59. Kim J, Taylor CR, Kim KH, Lee KH (2015) Measures of perceived sustainability. J Glob Scholars Market Sci 25(2):182–193
60. Stanton WJ, Futrell C (1987) Fundamentals of marketing. McGraw Hill, New York
61. Sun Y, Garrett TC, Kim KH (2016) Do Confucian principles enhance sustainable marketing and customer equity? J Bus Res 69(9):3772–3779
62. Rudawska E (2019) Sustainable marketing strategy in food and drink industry: a comparative analysis of B2B and B2C SMEs operating in Europe. J Bus Ind Mark 34(4):875–890
63. Sun Y, Ko E (2016) Influence of sustainable marketing activities on customer equity. J Glob Scholars Market Sci 26(3):270–283
64. Siggelkow N (2007) Persuasion with case studies. Acad Manage J 50(1):20–24
65. Tsang EW (2013) Case study methodology: causal explanation, contextualization, and theorizing. J Int Manag 19(2):195–202
66. Chreim S, Williams BE, Hinings CR (2007) Interlevel influences on the reconstruction of professional role identity. Acad Manage J 50(6):1515–1539
67. Coviello NE, Joseph RM (2012) Creating major innovations with customers: insights from small and young technology firms. J Mark 76(6):87–104
68. Dubois A, Gadde LE (2002) Systematic combining: an abductive approach to case research. J Bus Res 55(7):553–560
69. Becattini G (2000) Il bruco e la farfalla. Prato nel mondo che cambia (1954–1993). Le Monnier, Firenze
70. Guercini S, Milanesi M, Dei Ottati G (2017) Paths of evolution for the Chinese migrant entrepreneurship: a multiple case analysis in Italy. J Int Entrep 15:266–294
71. Piscitello L, Sgobbi F (2004) Globalisation, e-business and SMEs: evidence from the Italian district of Prato. Small Bus Econ 22:333–347
72. Milanesi M, Guercini S, Waluszewski A (2016) A black swan in the district? An IMP perspective on immigrant entrepreneurship and changes in industrial districts. IMP J 10(2):243–259

73. Creswell JW, Miller DL (2000) Determining validity in qualitative inquiry. Theory Pract 39(3):124–130
74. Woodside AG, Wilson EJ (2003) Case study research methods for theory building. J Bus Ind Mark 18(6/7):493–508
75. Eisenhardt KM, Graebner ME (2007) Theory building from case study research. Acad Manage J 50(1):25–32
76. Varadarajan R (2017) Innovating for sustainability: a framework for sustainable innovations and a model of sustainable innovations orientation. J Acad Mark Sci 45(1):14–36
77. Maignan I, Ferrell OC (2001) Corporate citizenship as a marketing instrument-concepts, evidence and research directions. Eur J Mark 35(3/4):457–484

Renewable Energy Importance and Possible Applications in Textile

Sanjoy Debnath, Manisha Jagadale, Naveen Jose, and Mahesh Jadhav

Abstract The textile industry is a major consumer of water and energy, with substantial reliance on fossil fuels, resulting in significant environmental challenges. The finite availability and unsustainable nature of fossil fuels, combined with their contribution to environmental degradation, underscore the urgency of transitioning to alternative energy sources. This chapter examines the energy-intensive operations of the textile industry, including fiber production, yarn manufacturing, dyeing, finishing, and waste management, and explores the integration of renewable energy solutions such as solar, wind, biomass, and geothermal technologies. Through a comprehensive analysis, the chapter also highlights case studies and best practices that demonstrate the feasibility and benefits of adopting renewable energy. Key focus areas include reducing carbon emissions, lowering operational costs, and enhancing environmental sustainability. Economic and technological considerations, including cost-effectiveness, scalability, and compatibility with existing systems, are also discussed to address challenges in the transition process. The insights aim to provide a roadmap for industry stakeholders to adopt sustainable energy practices, paving the way for a greener and more resilient textile production model. The insights provided aim to guide industry stakeholders toward a greener, more sustainable textile production model.

Keywords Challenges · Energy · Environment · Sustainable textile · Renewable energy · Textile industry

S. Debnath (✉) · M. Jagadale · N. Jose
ICAR—National Institute of Natural Fibre Engineering and Technology, Kolkata, West Bengal, India

M. Jadhav
Dr. Balasaheb Sawant Konkan Krishi Vidyapeeth, Dapoli, Maharashtra, India

S. S. Muthu (ed.), *Renewable Energy Implications for the Textile Industry*, Sustainable Textiles: Production, Processing, Manufacturing & Chemistry, https://doi.org/10.1007/978-3-032-10917-0_3

1 Introduction

The escalating global energy crisis, largely attributed to the relentless exploitation of non-renewable fossil fuels such as coal, crude oil, and natural gas, has led to severe environmental degradation and socio-economic instability. The unchecked extraction and combustion of these fuels have resulted in multiple adverse consequences including deforestation, soil erosion, water pollution, and most significantly, a substantial surge in greenhouse gas (GHG) emissions [1]. These emissions are a primary driver of anthropogenic climate change, intensifying the warming of the planet and contributing to erratic weather patterns, sea-level rise, and ecosystem disruption [2].

To address these pressing challenges, international policy instruments such as the Paris Agreement have been established, with the ambitious objective of restricting global temperature rise to well below 2 °C above pre-industrial levels. A key strategy in this effort involves the reduction of global carbon emissions, targeting a limit of approximately 40 billion tonnes annually by the year 2030 [3, 4]. In this context, the scientific and industrial communities have increasingly turned their focus toward identifying and harnessing sustainable and renewable sources of energy that can ensure long-term energy security while minimizing environmental impact.

The shift in focus toward renewable energy is particularly evident within the manufacturing sector, where the energy demand especially in the form of thermal energy has been rising. Modern industrial systems are increasingly being analysed from a holistic perspective that emphasizes energy efficiency, cost-effectiveness, and environmental responsibility. The dual pressures of mitigating climate change and reducing operational costs have driven a global push toward low-carbon technologies and the gradual decarbonization of energy-intensive sectors, such as textiles. According to projections by the International Energy Agency (IEA), global energy consumption is expected to increase by approximately 50% between 2018 and 2050 [5]. Global primary energy consumption by different sources in 2013 and 2023 given in Table 1. If current consumption patterns persist without significant shifts toward alternative energy sources, it is anticipated that the reserves of conventional fossil fuels particularly oil and natural gas may face critical depletion by around 2042. Such a scenario would likely trigger a severe energy crisis, emphasizing the urgent need for transitioning to sustainable and renewable energy systems [6, 7].

The textile and apparel manufacturing industry, in particular, is recognized as one of the most resource-intensive sectors in terms of energy and water consumption. Major operations such as ginning, spinning, weaving, dyeing, printing, and finishing consume substantial quantities of electricity and thermal energy [8]. While on a per-unit basis, textiles may not be the most energy-consuming products, the enormous scale of the industry especially the widespread proliferation of small and medium enterprises (SMEs) results in high cumulative energy demand. Apart from its energy footprint, the textile industry is also a significant source of both solid and liquid waste. The disposal of untreated effluents and residues, often rich in dyes, chemicals, and microfibers, places immense pressure on terrestrial and aquatic ecosystems, leading to the depletion of finite natural resources and the degradation of environmental quality [2, 9].

Amidst these environmental concerns, renewable energy emerges as a promising alternative to conventional fossil fuels [10]. Renewable sources such as solar, wind, hydro, geothermal, and biomass offer a path toward cleaner industrial energy practices [11]. Table 2 shows global renewable energy generation by different sources. These energy forms not only have minimal GHG emissions but also contribute to energy diversification, reduced dependency on volatile fossil fuel markets, and enhanced energy security [12, 13]. The deployment of such technologies within the textile sector is gaining momentum, driven by technological advancements, supportive policy frameworks, and growing awareness regarding corporate environmental responsibility.

Among these renewable technologies, solar energy is particularly well-suited for integration into textile manufacturing processes. Solar thermal systems can provide the requisite heat for water-intensive processes such as dyeing, bleaching, and drying, which typically require temperatures ranging between 60 and 150 °C). Simultaneously, photovoltaic (PV) systems are being utilized to generate electricity to power machinery, lighting, and other auxiliary systems in textile production

Table 1 Global primary energy consumption (terawatt-hours) by different sources in 2013 and 2023

Sources	2013	2023
Traditional biomass	11,330	11,111
Coal	44,709	45,565
Oil	49,654	54,564
Natural gas	33,721	40,102
Nuclear	6513	6824
Hydropower	10,324	11,014
Wind	1732	6040
Solar	378	4264
Modern biofuels	821	1318
Other renewables	1448	2428

Energy Institute—Statistical Review of World Energy [14]

Table 2 Global renewable energy generation by sources

	2022	2023
Wind	2108	2325.3
Solar	1321.9	1641.6
Hydro	4323.4	4240.0
Other renewables	776.8	781.5
Total	8530.1	8988.4

Energy Institute—Statistical Review of World Energy [14]

units. Regions with abundant solar irradiation such as India, Bangladesh, and various African nations are increasingly embracing solar energy for both thermal and electrical applications, enhancing operational sustainability.

Another viable renewable energy route lies in the utilization of biomass-based fuels. Biomass energy, derived from organic materials including agricultural residues and industrial by-products, holds significant potential in textile settings. Waste streams such as cotton linters, jute dust, kenaf residues, and other lignocellulosic materials can be harnessed to produce biogas, biodiesel, bioethanol, and even electricity through various conversion technologies including anaerobic digestion, pyrolysis, and gasification [15]. The classification of biomass into three generations first (edible sources like sugarcane), second (non-edible lignocellulosic biomass like cotton stalks or jute), and third (microalgal biomass) provides a structured framework for evaluating feedstock sustainability [16]. Given the environmental and ethical issues surrounding food-versus-fuel debates, current research is increasingly oriented toward second and third-generation biomass sources due to their higher sustainability indices and minimal competition with food production [17].

Significantly, the textile and fashion industries also produce massive amounts of post-consumer and in-process waste, much of which remains unutilized. These wastes primarily consist of natural fiber-based materials rich in cellulose, hemicellulose, and proteins components that are highly suitable for bioenergy generation. According to projections, global textile waste is expected to rise to 148 million tonnes annually by 2030, with over one-third of this waste being cellulose-based, offering a substantial but currently underexploited resource for renewable energy conversion [18]. Figure 1 shows fundamental operations involved in textile processing. Despite a growing body of literature on general waste valorization, research specifically focusing on energy recovery from textile-specific biowaste remains sparse. This gap represents a vital opportunity for innovation and interdisciplinary collaboration between material scientists, energy technologists, and environmental policymakers.

In addition to solar and biomass energy, wind and small-scale hydroelectric power also offer supplemental renewable energy options for the textile industry. These sources are particularly useful in remote or off-grid regions where stable grid connectivity is lacking. While these systems may involve higher initial infrastructure investments and location-specific feasibility assessments, their ability to provide decentralized, resilient, and low-emission energy makes them viable components of an integrated energy solution. Governmental policy instruments, international climate mandates, and private sustainability initiatives are all playing a pivotal role in accelerating the adoption of renewable energy in the textile sector. National strategies such as India's National Solar Mission, and global frameworks like the United Nations Sustainable Development Goals (SDGs), as well as voluntary corporate initiatives such as RE100, are helping catalyze the shift toward sustainable production. Textile manufacturers and fashion brands are increasingly investing in renewable energy technologies not only to meet environmental compliance standards but also to enhance brand value, reduce operational risks, and gain consumer trust in an environmentally conscious marketplace.

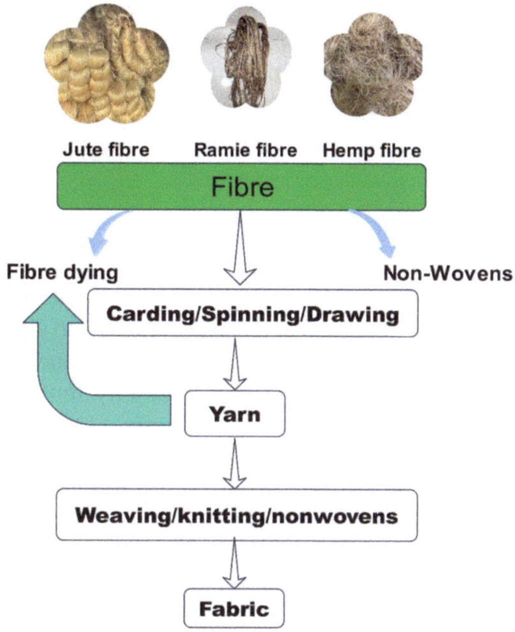

Fig. 1 Fundamental operations involved in textile processing

However, despite these positive trends, several persistent challenges continue to obstruct large-scale integration of renewables in textile manufacturing. These include high capital investment requirements, inadequate technical expertise among workers, compatibility issues with legacy machinery, and insufficient awareness of long-term benefits. Overcoming these hurdles will require an integrated approach involving policy support, public–private partnerships, financial subsidies, skill development programs, and dissemination of successful real-world case studies to build trust and capacity among industry stakeholders.

The primary objective of this book chapter is to study renewable energy application in textile industry. This book chapter also highlights the shortcomings of existing approaches, the gaps in current research, and the barriers to commercialization. This book chapter aims to inform researchers, policymakers, and industry leaders about feasible strategies for reducing carbon emissions while promoting energy resilience and sustainability in textile manufacturing.

2 Renewable Energy Sources and Their Applications

The global shift towards renewable energy (RE) has accelerated in recent years due to growing concerns about environmental degradation and climate change associated with conventional fossil fuel-based power generation. Renewable energy technologies are being adopted widely as they offer cleaner, sustainable alternatives.

The five primary categories of renewable energy solar, wind, biomass, hydropower, and geothermal each operate under specific environmental and technological conditions, and their efficiency and feasibility are largely influenced by geographical and climatic factors.

2.1 Solar Energy

Solar energy is harnessed directly from the sun and primarily utilized through two distinct methods: solar thermal systems and photovoltaic (PV) systems. Photovoltaic technology involves the use of semiconductor materials to absorb solar radiation and convert it directly into electricity through the photovoltaic effect. PV cells, commonly known as solar cells, are a key component in this process. In countries with abundant sunlight, such as India, decentralized solar PV systems like mini-grids are an efficient means of electrification, particularly in remote areas beyond the reach of conventional power infrastructure. India receives approximately 5000 trillion kWh of solar energy annually, with most regions experiencing solar insolation levels of 4–7 kWh/m^2/day. Globally, the installed capacity of solar PV reached 1865 GW by the end of 2024, a significant increase from 710 GW in 2020. In 2024 alone, 451 GW of solar PV capacity was added constituting the highest contribution among all renewable sources and accounting for over 75% of new renewable capacity installations [19].

2.2 Wind Energy

Wind energy is a mature and widely adopted form of renewable electricity generation due to its cost-effectiveness, scalability, and relatively simple infrastructure [20]. Wind turbines convert the kinetic energy of moving air into electrical energy. While the majority of wind farms are located onshore, offshore installations are expanding rapidly due to higher and more consistent wind speeds. The suitability of a location for wind energy generation depends on the annual average wind speed, which must be sufficient to ensure reliable power output. Elevated terrains and coastal regions are typically favorable. In India, the National Institute of Wind Energy (NIWE) has established over 900 wind-monitoring stations and developed wind potential maps at various hub heights (50 –150 m). The country's wind power potential is estimated at approximately 695.5 GW at 120 m and 1163.9 GW at 150 m above ground level. Most of this potential is concentrated in eight key states: Andhra Pradesh, Gujarat, Karnataka, Madhya Pradesh, Maharashtra, Rajasthan, Tamil Nadu, and Telangana [14].

2.3 Biomass Energy

Biomass energy is derived from organic materials and is categorized into traditional and modern applications. Traditional biomass includes the direct combustion of resources such as firewood, animal dung, and charcoal. Modern biomass energy encompasses more advanced technologies such as anaerobic digestion for biogas production, bio-refineries, pellet-based heating systems, and biofuels derived from agricultural residues like bagasse [17]. India generates an estimated 750 million metric tonnes of biomass annually, indicating significant potential for energy conversion [21]. The government has promoted co-firing of biomass (5–10%) in coal-based thermal power plants to enhance fuel sustainability. Biomass energy offers advantages such as decentralized power generation, carbon neutrality, renewability, and the potential to generate rural employment.

2.4 Hydropower Energy

Hydropower utilizes the kinetic energy of flowing or falling water to drive turbines for electricity generation. It is one of the most established and cost-effective renewable energy sources, with a conversion efficiency of up to 90% the highest among all RE technologies. Hydropower systems typically harness energy from rivers, streams, or artificial reservoirs with elevation differences. In India, hydropower accounts for approximately 11.5% of total electricity production [22]. The flexibility and scalability of hydropower systems from large dams to small run-of-river projects make it a versatile renewable source.

2.5 Geothermal Energy

Geothermal energy is derived from the Earth's internal heat, originating from geological processes and radioactive decay in the Earth's crust. The average geothermal gradient is about 30 °C per kilometer of depth. Certain regions exhibit significantly higher gradients, enabling the extraction of thermal energy for power generation and direct heating applications. Geothermal systems typically involve drilling wells into hydrothermal reservoirs, where hot water or steam is brought to the surface. These reservoirs are either naturally permeable (hydrothermal) or artificially enhanced (Enhanced Geothermal Systems, EGS). Depending on temperature and pressure, geothermal fluids can be utilized for electricity generation or industrial heat applications.

2.6 Ocean Energy

Ocean energy encompasses various marine-based energy forms including wave, tidal, thermal, and offshore wind energy. Surface waves are generated by wind interacting with the ocean surface; the energy carried by waves increases with wind speed, duration, and fetch. Oceans have vast potential to meet global energy demands many times over through their continuous and predictable energy cycles. Technologies to harness this energy include wave energy converters, tidal turbines, and ocean thermal energy conversion (OTEC) systems. Although still in developmental stages, ocean energy technologies offer promising prospects for sustainable, large-scale energy generation.

3 Energy Applications in Different Textile Operations

3.1 Energy in Conventional Textile Industry

It is well established that, today's textile industry involves mostly batch processes and to some extent continues process wherein without energy plays in important role. Consumption of energy in textile industry mainly in the form of electrical and thermal. In spinning and spinning preparatory, energy requirement is for opening, cleaning and mechanical processing fibre at different spinning machinery [23]. Before 2000, there were very less numbers of EOUs (export oriented units) in Indian textile industry and as a result age-old textile machines used in different spinning and weaving mills. Due to use of use of age-old technologies/machines the power consumption in Indian textile industry (spinning and weaving) is considerably high as discussed by [24]. Further, depending upon the product ranges, mainly for weaving of coloured sorts, there is a fall in efficiency of the processing machinery which again enhances the energy consumption [25]. The energy consumption/requirement in Most of the bast fibres are coarser in nature and because of their higher rigidity and toughness, requires more as a result twisting force during opening and spinning operations [26]. Jute industry is one such which still uses mostly old machinery, old drive system and more labour involvement is facing sustainability issues mostly in energy consumption aspects [27]. The electrical energy price is increasing day-by-day and the traditional industry (jute industry), the heavy moving components of the machinery requires higher energy. For instance, many jute industry are still using jute shuttle looms for sacking. These sacking looms used heave shuttle, requires more energy during pickup of shuttle and again to stop. The production speed is limited to maximum of 120 ppm (picks per minute). Except few, (Gloster Jute Mills, Birla Jute Mills, etc.) most of the jute mill owners, are not interested to introduce and invest on latest technologies. Most industries still using over 100 years old carding, drawing, spinning, etc., machineries. Even today, many of the mail use common shaft flat belt drive for carding. The slippage of flat belt drive is considerably higher than 'V-Belt' drive system. These causes more energy losses during conversion of fibre to yarn or yarn to fabric. It has been documented that for production of 1 kg of yarn spinning around 2.7 kWh of electrical energy was consumed in older machine processing [23]. However, in present scenario, with the introduction of more process control and additional devices/systems (air blowers, air suctions, automations, robotics applications, process control and automatic quality inspection etc.) the figure of electrical energy consumption is around 4 kWh. In addition, the thermal energy consumption per kg of textile material (yarn and fabric) processing (scouring, bleaching, dyeing, drying, finishing, etc.) increases from 1.1 MJ/kg in older system to 4.7 MJ/kg in modern system. In spinning mostly the air conditioning and filtration (dust/air filter) also significantly consume electrical energy.

3.2 Energy in New/Renovated Textile Industry

Document revealed that in modern staple fibre processing industry, out of total electrical power consumption, carding itself consumes around 9.44% [28]. Compared to old industry, illumination, temperature and humidity control, atmosphere cleanness system, compressor, etc. the total energy requirement is around 78%. As most of the modern machine requires pneumatic systems in different unit operations for effective and quality production. For instances, instead of spring loading in top arm of roller drafting in spinning, drawing, pneumatic system is used. Inspire of using different energy efficient motors/drives, many of the machines, uses different individual motors, instead of single motor; electric/electronic/photoelectronic, etc. sensors instead of mechanical type system; reduction of manual operation and replacement of such jobs by electronic systems, for instance, can changing mechanism, auto doffing, auto splicing; conveyors for automation in material shorting and transport system, etc. require additional energy in modern machinery in spinning and weaving sections. Earlier days power loom used for weaving, after renovation replaced by auto looms. Today most weaving are replaced by shutterless looms, such system uses more than one motors in each loom; pneumatic system is essential, consumes additional energy; air jet looms consumes good amount of energy in compresses and suction air generation; Open-end rotor spinning, two for one twister, automatic/pneumatic fibre mass transfer through chute feed system in blow-room, etc. consumes considerable amount of energy. In wet processing, with the stringent pollution norms, instead of disposing the textile effluent water in drainage, necessary processing/treatment of textile effluent/pollutant water also requires energy. In conclusion, modern textile energy consumes more energy compared to old textile mills. Modern textile industry, specially, wet processing industry, where heat energy is used, emits huge amount of greenhouse gases during burning of fossil fuel [1, 29]. This is again detrimental towards sustainable development goals. Hence, stress may be imposed on used of renewable sources of energy. An concluding analysis on distribution of energy in different textile operations are as follows:

Around 23% of the total energy consumed in weaving operation.
Around 34% of the total energy consumed in spinning operation.
Around 38% of the total energy consumed in chemical processing.
Around 5% of the total energy consumed in miscellaneous operations.

3.3 *Energy in Unconventional Textile Industry*

The unconventional textile industry considered as unconventional production line like natural fibre nonwoven. In this fabric is being prepared with minimum machinery operations/involvement directly from fibre and production system is continuous [30]. The natural fibre based nonwoven industry mostly using advanced machinery with introduction of energy efficient power transmission driving system, motors and most importantly improved alloy for high strength-light-weight moving machinery components. Energy requirement is considerable much less in such production industry. On the contrary, this system also requires compressor which consumes considerable electrical energy. Hydroentanglement, spun bonding, spun lace are some of the modern systems of nonwoven manufacturing also consumes significantly higher energy, however, due to higher production rate, the process are cost per unit is much effectively less. One of the good example of renewable is Gokak Textile mill. Gokak Textiles Limited (GTL) generates power through its subsidiary, Gokak Power & Energy Limited (GPEL), which owns and operates hydroelectric power projects at Gokak Falls. The total installed capacity of these hydroelectric units is 10.8 MW [31]. A Ludhiana based textile industry, Ganga Acrowools Limited, also in process of installing renewable power generation from agro-residue for own consumption which is under process of commissioning. Such initiatives are very much needed in todays' scenario towards reduction of carbon footprint and reduction dependency of fossil fuel.

4 Challenges of Renewable Energy

4.1 *Technical Feasibility and Integration Challenges*

Several technical challenges are often faced by organisations while adopting renewable energy sources. One of the most important among them is the intermittency of renewable sources like solar and wind energy, which makes it difficult to ensure a consistent energy supply for continuous industrial operations. Without reliable energy storage systems or efficient grid integration technologies, energy fluctuations can disrupt production. Technical expertise is also limited in installing, operating, and maintaining renewable energy systems, which further hinders their effective deployment. Many textile facilities, especially in developing regions, lack trained personnel who can manage the complexities of renewable technologies such as solar thermal units [1]. Another key issue is the compatibility of renewable systems with existing infrastructure. Retrofitting conventional boilers or integrating new energy systems into old manufacturing setups often requires significant technical modifications, which can be both costly and complex and also requires huge space for accommodating additional units [32]. The maintenance cost of the advanced technologies often falls on the higher side, deterring the medium and small-scale industries from adopting the same.

4.2 Financial Barriers

The benefits of the usage of renewable energy sources are often overshadowed by the immediate financial burdens faced by the small and medium-sized enterprises (SMEs), which form the backbone of the textile sector in many developing countries, for mobilizing the necessary capital for infrastructure [33]. Moreover, the payback period for renewable energy investments, although shorter in the long term due to savings on fuel and emissions-related costs, is often viewed as uncertain. Problems are further aggravated by limited credit access, lack of renewable energy-specific financial products and high perceived investment risk [34].

4.3 Regulatory and Policy Barriers

Outdated and inconsistent policies remain a major obstacle to renewable energy adoption in the textile sector. The lack of clear targets, supportive tariffs, and streamlined permitting processes reduces investor confidence and complicates project implementation. In many cases, regulatory delays and fragmented governance further discourage potential adopters. The government should try to address these challenges by introducing coherent and transparent policies and simplifying administrative procedures [35]. Engaging industry stakeholders in policy development can ensure practical and widely accepted regulatory frameworks, which can facilitate a smooth transition to clean energy.

4.4 Institutional and Market Resistance

While considering the textile industry, it was found that the short-term cost considerations outweigh long-term sustainability goals, which often creates an inherent resistance to widespread adoption of renewable energy sources [36]. Many industries continue to rely on fossil fuels due to their perceived reliability and affordability, despite having potential savings with the usage of renewable energy. There is a lack of familiarity or confidence among organizations to implement newer technologies, especially in the absence of internal incentives or policy mandates. To overcome this, fostering a culture of innovation, initiating pilot projects, and highlighting successful case studies can help shift perceptions [37].

5 Future Perspectives of Renewable Energy

The future of renewable energy in the textile industry is marked by transformative opportunities driven by technological advancements, evolving policy landscapes, and growing environmental consciousness. As the global textile industry faces growing concerns about its environmental impact, adopting renewable energy emerges as a promising solution for achieving sustainable production and ensuring energy security.

5.1 Advanced Energy Systems

The evolution of hybrid energy systems that synergise renewable sources with conventional energy to ensure reliability and efficiency is gaining popularity in the present era. Future research is expected to focus on intelligent energy management systems (EMS), which utilize real-time data analytics and predictive algorithms to optimize energy consumption in textile manufacturing units. The integration of Internet of Things (IoT) and artificial intelligence (AI) into hybrid setups can enable automated decision-making to switch between energy sources based on demand and supply conditions, thus mitigating issues of intermittency and load mismatches [38].

Development of advanced energy storage solutions like next-generation lithium-ion and solid-state batteries for wind and solar photovoltaic systems, high-capacity thermal storage systems for solar thermal energy, and scalable compressed air or hydrogen-based storage for grid support is essential to increase the reliability of renewable sources. Breakthroughs in materials science and battery chemistry are expected to reduce costs, improve cycle life, and increase energy density, making storage systems more accessible for medium- and large-scale textile operations [39].

5.2 Energy Optimization and Circular Economy Approach

Future systems may rely on high-fidelity simulations to predict the energy needs of different processes (e.g., dyeing, drying, spinning) and test renewable integration scenarios without physical trials. This predictive modelling will enhance system design, reduce downtime during retrofitting, and improve the overall efficiency of renewable energy usage in textile plants. Introduction of digital twin technologies, which create virtual replicas of physical textile production systems, is poised to revolutionize energy planning [40].

5.3 Policy Regulations and Incentives

On the policy and regulatory front, fragmented and inconsistent regulations often discourage investment. Establishing clear, streamlined policies, such as renewable energy mandates, fast-tracked permitting, and feed-in tariffs, can enhance regulatory transparency and build investor confidence. Active stakeholder engagement, involving industry, government, and community actors, ensures that policies are context-sensitive and ready for implementation. Financial barriers, particularly high upfront capital costs, can be mitigated through well-designed government interventions such as subsidies, tax incentives, and concessional loans. Development of green bonds, energy performance contracts, and public-private partnerships through green financing can bridge investment gaps. These mechanisms lower entry thresholds for industries, especially SMEs, facilitating broader adoption [33, 41].

5.4 Skill Development and Decentralized Energy Models

The future adoption of renewable energy technologies in the textile sector will also hinge on a skilled and adaptive workforce. Investments in technical training, certification programs, and curriculum integration of green energy concepts at vocational and tertiary education levels will be critical. Governments and industry bodies must collaborate to build human capital capable of designing, installing, maintaining, and innovating renewable energy systems tailored for textile applications [42].

A uniform energy model is impractical given the diversity in regional climates, infrastructure, and industrial demands. Instead, localized solutions tailored to regional strengths can significantly enhance the feasibility and sustainability of renewable integration. For instance, concentrated solar thermal systems can be efficiently employed in regions of high solar irradiance, especially in tropical and subtropical countries, for meeting the industrial heating needs in textile dyeing and processing units. A robust regional energy planning, including infrastructure assessment, demand forecasting, and resource mapping, is essential for the successful integration of clean energy sources.

5.5 Alignment with Sustainable Development Goals (SDGs)

The future trajectory of renewable energy adoption in textiles aligns with broader global goals, particularly the United Nations Sustainable Development Goals (SDGs). Specifically, Goal 7 (Affordable and Clean Energy), Goal 12 (Responsible

Consumption and Production), and Goal 13 (Climate Action) provide a foundational framework for industry stakeholders [43]. As international trade increasingly favours sustainable sourcing, textile manufacturers will be incentivized to demonstrate renewable energy usage as part of compliance and branding strategies. Life Cycle Assessments (LCAs) consistently demonstrate lower greenhouse gas emissions and a reduced environmental footprint for renewable systems compared to conventional energy sources [44].

Renewable energy adoption in the textile sector presents a pathway toward environmental sustainability, economic efficiency, and industrial resilience. While the challenges are complex, ranging from technical and financial constraints to regulatory hurdles, a collaborative and strategic approach involving the stakeholders can unlock the potential of the renewable energy sector to become a model of green industrial transformation. Installation of solar panels at Textile mill roof and open areas, installation of hydra-electric power plant wherever possible/feasible, installation wind mill is free space is available, use more transparent sheets at roof to tap the natural light and reduce the consumption of electrical energy on illumination/lighting during day time. Different government provides subsidy for installation of such renewable energy generation.

6 Conclusion

The introduction of renewable energy adoption in the textile industry towards environmental sustainability, economic efficiency, and industrial resilience is need of the hour. Utilization of agro-waste for energy generation through different pathway (gasification, bio oil, bio gas, etc.) will reduce the dependency on reserve fossil fuel. Tapping the natural sunlight for power generation at lease for the lighting/illumination in the textile mills will reduce a great extent of the energy consumption from convention sources. In this regard, also use more transparent sheets at roof to tap the natural light reduce the electrical energy consumption during day time. Different government provides subsidy for encouraging installation of such renewable energy generation. Use of more and more such renewable energy will lead towards achieving sustainable goals.

References

1. Farhana K, Kadirgama K, Mahamude ASF et al (2022) Energy consumption, environmental impact, and implementation of renewable energy resources in global textile industries: an overview towards circularity and sustainability. Mater Circ Econ 4:15. https://doi.org/10.1007/s42824-022-00059-1
2. Zong S (2023) The systematic application of plant dyeing techniques based on sustainable energy systems in ecological textile design. Clean Environ Syst 11:100143. https://doi.org/10.1016/j.cesys.2023.100143

3. Wang Y, Liu Y, Feng W, Zeng S (2023) Waste haven transfer and poverty-environment trap: evidence from EU. Green Low-Carbon Econ 1(1):41–49
4. Usman AM, Abdullah MK (2023) An assessment of building energy consumption characteristics using analytical energy and carbon footprint assessment model. Green Low-Carbon Econ 1(1):28–40
5. IEA (2021) World energy outlook 2021. International Energy Agency. https://www.iea.org/reports/world-energy-outlook-2021
6. Lior N (2008) Energy resources and use: the present situation and possible paths to the future. Energy 33(6):842–857. https://doi.org/10.1016/j.energy.2007.09.009
7. Shafiee S, Topal E (2009) When will fossil fuel reserves be diminished? Energy Policy 37(1):181–189. https://doi.org/10.1016/j.enpol.2008.08.016
8. Khandaker S, Bashar MM, Islam A, Hossain MT, Teo SH, Awual MR (2022) Sustainable energy generation from textile biowaste and its challenges: a comprehensive review. Renew Sustain Energy Rev 157:112051
9. Nunes LJR, Godina R, de Oliveira Matias C (2019) Technological innovation in biomass energy for the sustainable growth of textile industry. Sustainability 11(2):528. https://doi.org/10.3390/su11020528
10. Aravena AMO, Chen J, Brownbridge G, Akroyd J, Markus K (2025) An analysis of renewable energy resources and options for the energy transition in an analysis of renewable energy resources and options for the energy transition in Chile. Appl Energy 381:125107. https://doi.org/10.1016/j.apenergy.2024.125107
11. Hasanbeigi A, Price L (2012) A review of energy use and energy efficiency technologies for the textile industry. Renew Sustain Energy Rev 16(6):3648–3665. https://doi.org/10.1016/j.rser.2012.03.029
12. Hong GB, Su TL, Lee JD, Hsu TC, Chen HW (2010) Energy conservation potential in Taiwanese textile industry. Energy Policy 38(11):7048–7053
13. Teo SH, Huey CN, Islam A, Abdulkareem-Alsultan G, Joseph CG, Janaun J, Taufiq-Yap YH, Khandaker S, Islam GJ, Znad H, Awual MR (2022) Sustainable toxic dyes removal with advanced materials for clean water production: a comprehensive review. J Clean Prod 332:130039. https://doi.org/10.1016/j.jclepro.2021.130039
14. Ember. Energy Institute - Statistical Review of World Energy (2024) Available online: https://ourworldindata.org/energy. Accessed 25 June 2025.
15. Jagadale M, Gangil S, Jadhav M (2023) Enhancing fuel characteristics of jute sticks (Corchorus sp.) using fixed bed torrefaction process. Renew Energy 215:118992
16. Ghosh R, Bhattacherjee S (2013) A review study on anaerobic digesters with an insight to biogas production. Int J Eng Sci Invent 2(3):8–17
17. Jagadale M, Beula Isabel J, Jadhav M, Periyasamy S, Gizaw DG (2024) Biofuel production: past to present technologies. In: Inamuddin, Altalhi T (eds) Solid-gaseous biofuels production. Scrivener, Beverly, pp 1–30
18. Muneer T, Asif M, Cizmecioglu Z, Ozturk HK (2008) Prospects for solar water heating within Turkish textile industry. Renew Sustain Energy Rev 12(3):807–823
19. International Renewable Energy Agency (n.d.) Solar energy. IRENA. https://www.irena.org/solar. Accessed 7 July 2025
20. Ang TZ, Salem M, Kamarol M, Das HS, Nazari MA, Prabaharan N (2022) A comprehensive study of renewable energy sources: classifications, challenges and suggestions. Energy Strat Rev 43:100939
21. Bhuvaneshwari S, Hettiarachchi H (2019) Crop residue burning in India: policy challenges and potential solutions. Int J Environ Res Public Health 16(5):832. https://doi.org/10.3390/ijerph16050832
22. Government of India, Ministry of Power. https://powermin.gov.in/en/content/power-sector-glance-all-India. Accessed 20 Nov 2022
23. Anonymous (2025) Power consumption in spinning mills. https://www.textileschool.com/161/power-consumption-in-spinning-mills/. Accessed 31 May 2025

24. Debnath S (1993) Power consumption in spinning and weaving. Text Trends 36(9):41–44
25. Debnath S (1993) Weaving efficiency in colour sorts. Textile Trends 36(5):43–45
26. Debnath S (2017) Sustainable production of bast fibres (Chapter 3). In: Muthu SS (ed) Sustainable fibres and textiles, Textile institute book series. Elsevier, Duxford, pp 69–85. https://doi.org/10.1016/B978-0-08-102041-8.00003-2
27. Debnath S (2017) Sustainability in jute-based industries (Chapter 7). In: Muthu SS (ed) Sustainability in textile industry, Textile science and clothing technology. Springer, Singapore, pp 139–147. https://doi.org/10.1007/978-981-10-2639-3_7. (Published online on 15th October, 2016). Print ISBN 978-981-10-2638-6, online ISBN 978-981-10-2639-3
28. Anonymous (2025) Energy consumption of chosen spinning mill. https://www.textileschool.com/205/energy-consumption-of-chosen-spinning-mill/. Accessed 31 May 2025
29. Anonymous (2025) The textile industry's transition to renewable energy. https://www.fibre2fashion.com/industry-article/9496/the-textile-industry-s-transition-to-renewable-energy. Accessed 31 May 2025
30. Debnath S (2017) Sustainable production and application of natural fibre-based nonwoven (Chapter 13). In: Muthu SS (ed) Sustainable fibres and textiles. Elsevier, Duxford, pp 367–391. https://doi.org/10.1016/B978-0-08-102041-8.00013-5. Print ISBN 978-0-08-102041-8, online ISBN 978-0-08-102042-5
31. Anonymous (2025) https://gokaktextiles.com/about/#:~:text=The%20company%20has%20three%20power,a%20new%204.5%2DMW%20unit. Accessed 31 May 2025
32. Nair G, Verde L, Olofsson T (2022) A review on technical challenges and possibilities on energy efficient retrofit measures in heritage buildings. Energies 15(20):7472
33. Kumar L, Nadeem F, Sloan M, Restle-Steinert J, Deitch MJ, Ali Naqvi S, Sassanelli C (2022) Fostering green finance for sustainable development: a focus on textile and leather small medium enterprises in Pakistan. Sustainability 14(19):11908
34. Jones AW (2015) Perceived barriers and policy solutions in clean energy infrastructure investment. J Clean Prod 104:297–304
35. Soputra J (2025) Comparative analysis of national sustainable development strategies and their integration of green technology initiatives: a multi-regional policy examination. Trans Digit Soc Hum Behav Socioecon Stud 15(2):19–28
36. Batra G (2023) Renewable energy economics: achieving harmony between environmental protection and economic goals. Soc Sci Chron 2(2):1–32
37. Debnath B, Taha MR, Siraj MT, Jahin MF, Ovi SI, Bari AM, Raihan A (2024) A grey approach to assess the challenges to adopting sustainable production practices in the apparel manufacturing industry: implications for sustainability. Results Eng 22:102006
38. Nebey AH (2024) Recent advancement in demand side energy management system for optimal energy utilization. Energy Rep 11:5422–5435
39. Gür TM (2018) Review of electrical energy storage technologies, materials and systems: challenges and prospects for large-scale grid storage. Energ Environ Sci 11(10):2696–2767
40. Alam MD, Kabir G, Mirmohammadsadeghi S (2023) A digital twin framework development for apparel manufacturing industry. Decis Anal J 7:100252
41. Alharbi SS, Mamun A, Boubaker SM, Rizvi SKA (2023) Green finance and renewable energy: a worldwide evidence. Energy Econ 118:106499
42. Bibi S, Khan A, Fubing X, Jianfeng H, Hussain S (2024) Integrating digitalization, environmental innovations, and green energy supply to ensure green production in China's textile and fashion industry: environmental policy and laws optimization perspective. Environ Dev Sustain 1–41. https://doi.org/10.1007/s10668-024-05417-4
43. Solntsev A M, Akshalova R D (2021) Sustainable development goal 7 and sustainable development goal 13: possible ways of interactions. In B. Zavyalova;·Elena G. Popkova Industry 4.0: exploring the consequences of climate change, 23–32. Cham: Springer
44. Shamoushaki M, Koh SL (2024) Comparative life cycle assessment of integrated renewable energy-based power systems. Sci Total Environ 946:174239

Calorific Value, Ash Content and Moisture Cotton Stalk Bioenergy: A Renewable Approach for Sustainable Textile Manufacturing

Samar Gogari Radha

Abstract Cotton stalks, a by-product of cotton cultivation, are often left on the fields or burnt, contributing to environmental pollution and waste. This chapter examines the potential of cotton stalks as a renewable bioenergy source for sustainable textile production. With their high lignocellulosic content, cotton stalks can be efficiently converted into biofuels through processes such as pyrolysis, gasification, and anaerobic digestion. This biomass valorization not only provides a decentralized and cost-effective energy supply for rural textile operations but also addresses waste management and soil degradation concerns. The chapter details the physico-chemical properties of cotton stalks that make them suitable for energy production and explores sustainable technologies for their conversion. Integrating cotton stalk bioenergy into textile value chains reduces dependency on fossil fuels and promotes circular economy principles. Additionally, socioeconomic advantages like increased farmer income and rural electricity availability are examined alongside environmental advantages like reduced greenhouse gas emissions and better air quality. Successful implementation strategies are demonstrated by case studies from India and other cotton-producing countries, while techno-economic evaluations offer information on viability and the return on investment. Policy recommendations include subsidies for biomass energy projects, research support for advanced bioenergy systems, and awareness programs for farmers and textile stakeholders. The chapter concludes that leveraging cotton stalks for bioenergy not only strengthens the sustainability of the textile sector but also contributes meaningfully to climate resilience and rural development.

Keywords Renewable bioenergy · Sustainable textile production · Climate resilience · Cotton stalks · Valorization

S. G. Radha (✉)
School of Biological and Environmental Sciences, Shoolini University of Biotechnology and Management Sciences, Solan, Himachal Pradesh, India

S. S. Muthu (ed.), *Renewable Energy Implications for the Textile Industry*, Sustainable Textiles: Production, Processing, Manufacturing & Chemistry,
https://doi.org/10.1007/978-3-032-10917-0_4

1 Introduction

1.1 Cotton Cultivation and Associated Agricultural Waste

Cotton (*Gossypium* sp.) being one of the most important cash (commercial) crops, is widely cultivated across diverse agro-climatic regions, particularly in countries like China, India, U.S. and Brazil. Even though the primary product of cotton cultivation is the fiber used in the textile sector, it generates significant amounts of agricultural waste, mainly being cotton stalks, roots leaves and bolls, which are left on the farm after harvest [1–3]. Several tons of lignocellulosic residue (majorly cotton stalks) are produced for every ton of cotton fiber produced, with estimates ranging from 10 to 15 tons ha^{-1} depending on regional yields and agronomic practices. The major off-farm and on-farm by-products are gin trash and cotton stalks, respectively. Traditionally, these residues are utilized as low-grade fuels or burned on the field itself to make the land clear for subsequent cultivation. Open burning, however results in polluting the environment, by emitting greenhouse gases (GHGs) and depleting of essential plant nutrients [1, 4]. These by-products which are usually regarded as waste, have a tremendous potential for valorization in various industrial and agricultural applications [3]. Exploring innovative ways of utilizing cotton waste is essential since integrating sustainable methods of agriculture into cotton cultivation can reduce environmental effects and improve resource efficiency [5]. In contrast, despite the rising emphasis on sustainable practices, many regions continue to rely on traditional conventional methods that do not adequately address the environmental problems posed by cotton production.

This chapter explores the transformation of cotton stalks, an abundant agricultural residue, into renewable bioenergy in various textile manufacturing processes. It provides details of physicochemical properties, bio-conversion technologies (like pyrolysis and anaerobic digestion), and integration models for biomass energy in textile clusters. Environmental, economic, and policy dimensions are discussed through case studies, emphasizing the role of cotton waste in achieving climate resilience and circular economy in textiles sector.

1.2 Environmental Issues Related to Cotton Stalks Burning

Burning cotton stalks residues causes major environmental issues primarily GHG emissions and release of other air pollutants. The combustion releases significant quantities of carbon dioxide (CO_2), nitrogen oxides (NO_x), carbon monoxide (CO), methane (CH_4) and particulate matter (PM), deteriorating the air quality and further exacerbating the problem of global warming [6, 7]. Approximately 0.85 million metric tons of CO_2 equivalent (CO_2e), are released when 1 million ton of cotton stalks are combusted, thus indicating its substantial carbon footprint [7]. The toxic chemicals and minute PM particles present in the smoke cause air pollution which is indirectly associated with cardiovascular and respiratory health problems in humans, particularly in rural areas where stubble burning is practised on a large scale [6, 8]. This pollution also deteriorates urban air quality by causing smog (smoke and fog) formation and thereby reducing visibility. Furthermore, burning destroys the structure of soil as well as its fertility and biodiversity in agricultural lands by destroying important soil nutrients and organic carbon [6, 9]. Alternative uses of these residues like converting cotton stalks into bioenergy, industrial raw materials or biochar can alleviate the environmental harm caused by burning them. This preserves soil health, lowers emissions and benefits farmers financially [7, 8]. Additionally these sustainable approaches also have the potential to generate carbon credits, which would encourage reduced burning and stimulate cleaner agricultural waste management [7].

1.3 Bioenergy as a Sustainable Solution in Textiles

Bioenergy derived from textile waste offers a sustainable solution to the energy and environmental challenges faced by textile industries. The cellulose-rich textile biowaste, generated at different processing and post-consumer stages, particularly cotton-based waste which contains 88–96% cellulose, thus making it a potential feedstock for production of bioenergy like biogas and bioethanol [10]. Enzymatic hydrolysis combined with pretreatment methods like NaOH or HPAC-NaOH (Hydrogen Peroxide Acetic Acid–Sodium Hydroxide) considerably improve the conversion efficiency of cotton and blended cotton-PET textile waste into fermentable sugars, which can then be further converted into bioethanol with yields exceeding conventional lignocellulosic biomass [10]. It has also been demonstrated that catalyst-supported pyrolysis of textile waste yields value-added energy carriers, providing a thermochemical pathway for bioenergy generation along with other biochemical methods [11]. Likewise, anaerobic digestion of pretreated textile fibers exhibits potential for biogas production, which would contribute to generate sustainable and renewable energy while efficiently managing textile waste [12]. Incorporating bioenergy production into textile waste management supports the principles of circular economy by converting waste into renewable fuels and platform chemicals like lactic acid, ethylene glycol and sorbitol that can be utilized in the textile manufacturing or other industries [10]. Aligning with sustainability objectives in the textile sector this approach not only reduces reliability on fossil fuels but also mitigates environmental pollution caused by textile waste disposal [10, 13].

1.4 Study Objectives and Implication in the Renewable Energy Context

This study aims to shed light on the valorization potential of cotton stalks as a renewable energy source within the textile industry value chain, with an emphasis on its capacity in reduction of environmental impacts and encourage circular economy approaches. This study investigates how cotton agricultural and textile waste can be successfully converted into renewable and clean bioenergy through thermochemical and biochemical conversion methods to meet the significant energy demands in textile manufacturing processes, particularly during dyeing, washing, and finishing. By emphasizing cotton waste as a feasible alternative to fossil fuels, the study makes a significant contribution to the sustainable transformation of the textile sector. It accentuates how textile clusters can lower GHG emissions by integrating cotton biomass-derived bioenergy, thus reducing the reliance on non-renewable energy and enhance energy efficiency. The study also discusses effective implementation methods like supply chain logistics, decentralized energy systems and policy recommendations. By aligning bioenergy initiatives with textile sustainability goals, this study contributes to widespread and broad environmental as well as socioeconomic objectives of waste reduction, carbon neutrality, clean energy access and rural income substitute. This positions renewable energy in the global textile value chain, as a transformative solution.

2 Physicochemical Properties of Cotton Stalks

2.1 Lignocellulosic Composition

Cotton stalks, a major agricultural residue, exhibit a lignocellulosic composition predominated by cellulose (32–46%), hemicellulose (20–28%) and lignin (15–30%), with some deviations based on the processing techniques and measurement methods [9]. Standard analyses by employing sulphuric acid hydrolysis and colorimetric methods reveal average values of 45% cellulose, 20% hemicellulose and 21% lignin, along with negligible ash content [14]. The cellulose fraction primarily comprises of crystalline microfibrils surrounded in a matrix of hemicellulose and lignin, creating a recalcitrant structure which is resistant to enzymatic degradation [15]. Cellulose content in untreated cotton stalk bark is ~41%, but steam flash explosion at 2.5 MPa further increases it to 72% by removing hemicellulose and lignin [16]. This aligns with studies showing thermal stability of cellulose (up to 350 °C) and it's potential for bioenergy applications or fiber extraction [15, 16]. Hemicellulose, a heteropolymer of xylose and arabinose, hydrolyzes more easily than cellulose, but its covalent bonds with lignin makes separation complicated [9, 15]. The complex phenylpropane, polymer Lignin, limits the efficiency of bioconversion by acting as a structural barrier [9, 14]. Lignin degradation by white-rot fungi (*Pleurotus ostreatus*) exhibits strain-dependent selectivity. These highly efficient strains of white-rot fungi can degrade lignin at rates reaching 14.82% within 15 days and up to 45.04% after 30 days. Nuclear magnetic resonance (NMR) studies reveal the presence of guaiacyl (G) and syringyl (S) units in lignin, with observed shifts in the G:S ratio during fungal treatment, indicating a selective degradation of S-units [14]. The interchange of these components impacts potential applications, wherein high cellulose content supports fiber-reinforced complexes, while recalcitrance of lignin requires pretreatment (e.g., steam explosion, fungal delignification) for biofuel production [14, 16]. This variability in reported compositions stresses the need for standardized characterization procedures to enhance valorization strategies [9, 15]. Table 1 summarizes the physiochemical properties of cotton stalks.

2.2 Calorific Value, Ash Content and Moisture Content of Cotton Stalks

According to some studies, the calorific value of cotton stalks ranges between 14.5 and 19.2 MJ/kg (typically 15.2–16.8 MJ/kg), indicating good potential for bioenergy compared to other biomass fuels. Bomb calorimetry revealed that cotton stalks have a calorific value of about 15.2 MJ/kg, indicating their possible utilization in combustion and gasification applications [18]. Despite variations, moisture content in cotton stalks is generally low, but sufficient for boosting the efficiency of combustion. Experimental data, in some studies indicate moisture levels to be ranging between 6.8 and 13.6% (dry basis), while in some it was documented as 6.84% and marginally higher values (~13.3%), signifying that optimal energy conversion needs drying [17, 18]. Ash content in cotton stalks is characteristically low ranging between 3 and 5%. In some studies, the ash content was found to be approximately 3.28%, while other proximate analyses found ash around 4.95%. Hence, compared to several other agricultural residues, cotton stalks produce significant amount of ash [17, 18]. Therefore all these qualities make cotton stalks a potential renewable biomass resource for energy generation, particularly in the form of briquettes or pellets. This enables sustainable use of agricultural residues and reduce environmental pollution resulting from open-field burning. [9, 17].

2.3 Comparative Overview of Cotton Stalks with Other Agricultural Residues

Cotton stalks exhibit beneficiary qualities compared to other agricultural residues when it comes to bioenergy applications. Their calorific value aligning with sugarcane bagasse (16.8–18.9 MJ/kg) (15.2–17.7 MJ/kg), exceeds rice husk (14.3–25.4 MJ/kg) and rice straw (minimal energy due to high moisture content) [9,

Table 1 Physicochemical properties of cotton stalks

Property	Reported range/value	References
Cellulose content (%)	32–46%; Avg. ~45%; up to 72% (after pretreatment)	[9, 14, 16]
Hemicellulose content (%)	20–28%; Avg. ~20–27%	[9, 15]
Lignin content (%)	15–30%; Avg. ~21%	[9, 14]
Ash content (%)	3.0–5.0%; e.g., 3.28%, 4.95%	[17, 18]
Moisture content (%)	6.8–13.6% (dry basis)	[17, 18]
Calorific value (MJ/kg)	14.5–19.2; typical ~15.2–16.8	[17, 18]

18, 19]. However, compared to rice husk (17.2–18.6%) and rice straw (15.2%), cotton stalks have a substantially lower ash content (3.28–6.93%), which reduces the risk of slagging during combustion. Although sugarcane bagasse has a similar low ash content (3.28%), but its energy efficiency is complicated because of higher moisture content (12–14%) compared to cotton stalks (6.8–11.5%) [18, 20, 21]. Moisture content in cotton stalks (6.8–13.6%) is lower compared to rice straw (12.2%) and other wetland herbs, indicating better combustion efficiency [18, 19]. Structurally, the lignin content (15–30%) of cotton stalks aligns with woody biomass providing flexibility for thermochemical processing, while the holocellulose content (67–70%) exceeds many other crop residues, allowing efficient conversion to biofuels [9, 15]. Sugarcane bagasse, despite having lower ash content, requires pre-processing because of its fibrous structure, unlike cotton stalks which have adaptability and flexibility to pelletization (bulk density: 146.6–485.4 kg/m^3) [20, 21]. Cotton stalks are therefore superior to rice residues and competitive with bagasse for sustainable bioenergy systems because they balance desirable energy density, manageable ash content and moderate levels of moisture. Table 2 gives a comparative overview of cotton stalks with other agricultural residues.

2.4 *Suitability of Cotton Stalks for Energy Conversion*

Cotton stalks demonstrate strong suitability for energy conversion through pyrolysis and gasification, offering a sustainable alternative to open-field burning. Their calorific value (15.2–21.9 MJ/kg) aligns with hardwood and exceeds many agricultural

Table 2 Comparison of cotton stalks with other agricultural residues

Parameter	Cotton stalks	Rice husk	Rice straw	Sugarcane bagasse	References
Calorific value (MJ/kg)	15.2–17.7	14.3–25.4	Lower (moisture sensitive)	16.8–18.9	[9, 18, 19]
Ash content (%)	3.28–6.93	17.2–18.6	~15.2	~3.28	[18, 20, 21]
Moisture content (%)	6.8–13.6	—	~12.2	12–14	[19, 20]
Holocellulose content (%)	67–70	—	—	—	[9, 15]
Lignin content (%)	15–30	—	—	~15–25	[9, 15]
Silica/ash composition	Low	High (up to 18%)	High (~15%)	Low	[19, 20]
Bulk density (kg/m^3)	146.6–485.4	—	—	—	[20, 21]
Pelletization feasibility	High	Moderate	Moderate	Requires pre-processing	[20, 21]

residues, making them viable for thermochemical processes [22, 23]. Environmental and Economic Benefits: Utilizing cotton stalks for energy reduces open-field burning emissions, mitigating 0.85 million metric tonnes of CO_2-equivalent per million tonnes burned [22, 24]. Cotton stalks' low ash (3.28–4.95%) and moderate moisture (6.8–13.6%), minimize slagging risks and preprocessing needs [23, 25]. Integrated systems can yield carbon credits, while biochar and syngas production supports circular economy goals, enhancing rural energy security [24, 25]. Cotton stalks are pyrolyzed at 300–800 °C to produce syngas biochar and bio-oil. Higher temperatures (600–800 °C) favor the production of bio-oil (16.2–18.9 MJ/kg) and syngas. Temperature of 400 °C yields highest amount of biochar (46.5%) with energy densification ratios up to 1.41 [25, 26]. The process is energy-efficient, as non-condensable gases (NCG) from pyrolysis can be combusted to provide 123 kW of surplus energy for plant processes [26]. An advantage of pyrolysis is its focus on production of solid/liquid fuels, with biochar (25.8 MJ/kg at 600 °C) suitable for combustion or soil improvement [25]. Compared to open burning, air and steam gasification generate syngas that is rich in H_2 (up to 39% v/v), CH_4, and CO along with lower CO_2 emissions [27]. Steam gasification enhances lower heating value (LHV) of syngas compared to air, facilitating use in thermal systems or engines [22, 27]. In comparison with direct combustion, gasification reduces particulate emissions by 90% and enables production of gaseous fuel [23].

3 Conversion Technologies for Bioenergy Production

3.1 Thermochemical Methods

Thermochemical conversion technologies are essential for converting biomass into bioenergy, thus providing fossil fuel alternatives which are sustainable. Combustion, pyrolysis, and gasification are key methods differing in operating parameters, products, and applications, each playing a distinct role in renewable energy systems. Figure 1 shows the diagrammatic representation of various conversion methods for bioenergy production from cotton stalks.

3.1.1 Pyrolysis

At 300–800 °C in anoxic conditions, pyrolysis thermally breaks down biomass to produce syngas (gas), bio-oil (liquid) and biochar (solid). Although the high oxygen content (~35–40%) of the oil requires upgrading for fuel applications, fast pyrolysis at 400–600 °C and residence time of 2–10 s, increases bio-oil production (up to 75 wt %) [29]. Slow pyrolysis prioritizes biochar, which improves soil fertility and carbon sequestration. In order to achieve net-positive energy outputs, integrated

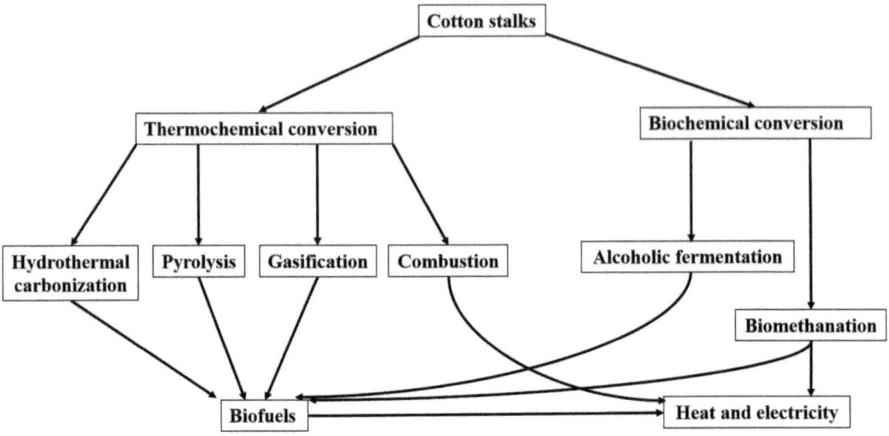

Fig. 1 Conversion methods for bioenergy production from cotton stalks. (Source: Al Afif et al. [28])

systems use NCG to balance energy demands [30]. Challenges in this method include feedstock variability and the need for efficient separation of products.

3.1.2 Gasification

By using sub-stoichiometric oxygen or steam, gasification converts biomass into syngas (primarily H_2, CO, CH_4) at 650 °C–1200 °C. While entrained-flow gasifiers (>1000 °C) achieve near-complete carbon conversion (>99%) and tar-free output, fluidized-bed gasifiers (650 °C–950 °C) handle a variety of feedstocks and produce syngas with a lower heating value (LHV) of 4–6 MJ/m^3 [29]. Steam gasification increases the hydrogen content (up to 48–52%), making it appropriate for hydrogen production or synthetic fuels. Compared to open burning, gasification reduces PM emissions by 90%, aligning with decarbonization objectives [31].

3.1.3 Combustion

For generation of heat and electricity, combustion involves oxidizing biomass at 800 °C–1000 °C. Being the most well-established method, combustion has efficiencies of 20–40% for generation of electricity when combined with steam turbines [29]. Dry, low-ash biomass like wood chips or agricultural residues can be combusted directly. However, PM, NO_x and CO_2 are released in incomplete combustion, imposing the need for advanced filtration systems [31]. Modern systems that optimize air supply and temperature control, including fluidized-bed combustors, improve efficiency and lower emissions.

3.1.4 Comparative Evaluation

When combined with carbon capture, pyrolysis and gasification provide pathways for carbon-negative energy whereas combustion is economical but has certain challenges in controlling emission. Pyrolysis offers bio-oil as an alternative transport fuel, though upgradation of costs between $0.50 and $1.50 per liter hinder its commercialization. Pyrolysis-derived biochar improves soil carbon storage, while gasification supports carbon credits and hydrogen economies [30]. Hybrid systems, like torrefaction pre-treatment followed by gasification, maximize energy efficiency and lower the risk of biomass ash slagging risks. Cotton stalks represent as a perfect feedstock for these processes because of their high cellulose content (32–46%) and low ash content (3–5%), which help mitigate emissions from open-field burning and promotes rural energy security [31].

3.2 Biochemical Methods

Biochemical conversion technologies are essential for transforming organic materials into bioenergy, thus providing fossil fuel alternatives which are sustainable. Two prominent methods that use microbial processes to transform biomass into renewable energy are anaerobic digestion for biogas and fermentation for bioethanol. These technologies help achieve the circular bioeconomy goals while addressing waste management challenges.

3.2.1 Anaerobic Digestion

Anaerobic digestion (AD) is a multi-stage biochemical process in which complex organic matter is broken down by a consortium of anaerobic microorganisms in anoxic conditions. This process produces a digestate that is rich in nutrients and biogas (mainly CH_4 and CO_2) [32–34]. Hydrolysis, acidogenesis, acetogenesis, and methanogenesis are the four main stages involved in the process. High-molecular-weight compounds (complex polymers) including carbohydrates, lipids and proteins are hydrolysed to produce simple molecules like sugars, fatty acids and amino acids. These molecules are further converted by acidogenesis into alcohols, carbon dioxide, hydrogen and volatile fatty acids (VFAs). These (VFAs) intermediates undergo acetogenesis which produces acetic acid, hydrogen, and carbon dioxide. Methanogenesis is the final stage in which methanogenic archaea produce methane and carbon dioxide from acetic acid and hydrogen/CO_2. Methanogenic archaea use hydrogen and acetic acid to produce 50–75% methane and 25–50% carbon dioxide [29]. Factors including substrate composition, pH (preferably 6.7–7.5) [35], temperature (mesophilic: 35–40 °C; thermophilic: 50–60 °C) [25], carbon-to-nitrogen ratio (16:1–25:1) and hydraulic retention time (usually 10–30 days), all decide the efficiency of AD [32, 34]. Potential feedstocks include lignocellulosic residues like rice straw, cotton stalks and waste textiles; however due to their recalcitrant structure they frequently need pre-treatments either mechanical, chemical, or biological in order to improve their biodegradability [32, 34]. AD is recognized for its low energy input, minimum sludge formation, and the simultaneous benefits of producing biofertilizer and renewable energy generation [32]. The process is widely adopted in agricultural biogas plants and municipal solid waste management, which boosts rural economies and contributes to job creation. AD reduces GHG emissions by diverting waste from landfills. For instance, pre-treatment biomass increases methane production by 20%, while digestion of cotton stalks can yield 65–86 L CH_4/kg volatile solids [25]. Integrated systems utilizing digestate for biochar production combine AD with pyrolysis for enhancing energy recovery by 42% [35].

3.2.2 Fermentation

Fermentation is the biological process that converts sugars into ethanol and CO_2, primarily using yeast (*Saccharomyces cerevisiae*) or specific bacteria [36–38]. This process can utilize a variety of feedstocks based on their biochemical composition. Sugary crops like sugarcane and sugar beet undergo direct fermentation after extraction of juice. Starchy crops like corn and wheat require enzymatic hydrolysis (particularly liquefaction and saccharification) to convert starch into fermentable sugars before undergoing the process of fermentation [38]. More complex feedstocks like lignocellulosic biomass including rice straw and cotton stalks, require pre-treatment using acid, alkali or enzymatic methods to release fermentable sugars from cellulose and hemicellulose, followed by fermentation [36, 37]. Usually industrial fermentation is carried out in continuous, batch or fed-batch modes. While continuous fermentation provides higher productivity and cost efficiency, the Melle-Boinot process remains a conventional batch method. For optimal yeast activity, the pH is typically maintained around 6.0 [38]. There is ongoing research on lignocellulosic feedstocks to improve enzyme efficiency and overcome inhibitors (like furans and phenolics from pre-treatment) [36, 37]. Syngas fermentation is an emerging method for production of bioethanol from gases derived from biomass. This method combines gasification and microbial conversion, [37]. Energy security, GHG mitigation and rural employment are all aided by the production of bioethanol. Owing to all this bioethanol is extensively used as a transportation fuel, either separately or as a blend with gasoline [36].

4 Application in the Textile Value Chain

4.1 Energy Use in Textile Processing

Textile processing uses a lot of energy, and both electrical as well as thermal energy are important at different stages. While wet processing steps like sizing, dyeing, washing and finishing demand substantial thermal energy, mostly in the form of steam for heating water and drying fabric; spinning and weaving processes rely mostly on electricity [39–41]. Wet processing is the most energy-intensive part of knitted fabric facilities, with studies indicating that it can account for up to 69% of steam consumption [42]. Thermal energy is majorly used in composite mills for heating and drying processes, and it is delivered by steam via extensive piping networks. [39]. Natural gas often dominates and accounts for up to 98% of total energy consumption in certain facilities, with electricity and diesel making up the remaining portion. The energy mix varies by location and facility. Seasonal variations exist in specific energy usage, which can range from 16.63 to 28.59 GJ per ton of fabric produced [43]. About 25% of the energy used to produce a final product comes from dyeing and finishing activities, which use a lot more energy than dry processes like knitting or weaving. Finishing procedures can use up to 17.9 kWh/kg [41]. Vacuum impregnation, effective drying technologies, and alternative heating techniques like radio frequency and heat transfer fluids are examples of energy-saving techniques that can result in tremendous energy use reductions [39, 41]. Sustainability is further improved by incorporating renewable energy sources in process improvements like supercritical CO_2 dyeing, which reduce energy usage and environmental effect [40, 41].

4.2 Replacing Fossil Fuels with Biomass-Derived Energy

Replacing fossil fuels with biomass-derived energy, particularly from agricultural residues such as cotton waste, presents a promising pathway for decarbonizing the textile value chain. Research demonstrates that cotton waste (a byproduct of textile and apparel manufacturing), can be effectively utilized in biomass boilers to generate process heat and steam, directly substituting fossil fuels and firewood. The calorific value of cotton waste aligns with conventional biomass fuels, ensuring technical feasibility for use in current boiler systems without major modifications [44]. Feasibility studies show several key benefits of utilizing cotton waste as a fuel source. Environmentally this approach contributes to waste diversion from landfills and lowering GHG emissions. Financially, it offers cost-efficiency through reduced fuel and waste disposal expenses. Moreover this practice aligns with legal and regulatory frameworks in several textile-producing regions, if proper guidelines are followed [44]. Additionally, experimental work shows that blending textile waste with other biomass types can produce low-emission but high-calorie fuel pellets suitable for industrial heating applications [45]. Apart from direct combustion, cotton waste can be converted into bio-oil or biogas through pyrolysis or anaerobic digestion, further expanding its utility as a renewable energy source within the textile sector [2, 13]. These approaches collectively support a circular economy, enhance resource efficiency, and contribute to the industry's transition toward sustainable energy systems.

4.3 Practical Models for Implementation in Textile Clusters

Practical models for implementing biomass-derived energy in textile clusters focus on integrating local agricultural residues-such as cotton waste-into centralized or distributed energy systems for process heating and steam generation. One effective model is the Build-Own-Operate (BOO) approach, where a third-party energy service provider installs and manages biomass-fired boilers within textile parks or clusters, supplying steam or hot water to multiple factories. This model reduces upfront investment for textile manufacturers and ensures professional operation and maintenance, as seen in successful case studies in India [46]. Another model involves establishing localized supply chains for collecting, processing, and transporting cotton and other agricultural residues to cluster-based biomass energy plants. These supply chains require robust monitoring, reporting, and verification systems to ensure sustainability and avoid negative land-use impacts [47]. Financial feasibility is enhanced by the reduced cost of steam production and decreased reliance on external fuel suppliers, as demonstrated in Sri Lankan garment factories where unit costs of steam dropped significantly after switching to cotton waste [44]. Technical feasibility is supported by the compatibility of cotton waste with standard biomass boilers, requiring minimal modifications. Legal and environmental compliance is ensured through adherence to local regulations regarding waste management and emissions. These models, when combined with energy audits and continuous optimization, enable textile clusters to achieve substantial cost savings, lower carbon emissions, and foster a circular economy by valorizing textile and agricultural waste streams [44, 47].

4.4 Storage, Transport, and Handling of Cotton Stalk Biomass

Because of its low bulk density, seasonal availability and vulnerability for deterioration, cotton stalk biomass presents special logistical and technological challenges for handling, storage, and transportation. Cotton stalks are usually stacked or piled in open fields, which results in substantial losses of 20–30% because of bugs, fungus, and moisture. These losses can be decreased to 10–15% with better storage techniques like thatched buildings or tarpaulin covering. Cotton stalks are usually processed into briquettes or pellets for effective handling and long-term storage. This reduces storage capacity by up to 80% and increases shelf life to 6–12 months by keeping moisture content below 12%. [48]. Typically, the collection procedure involves uprooting (either manually or mechanically) and then shredding or chipping to make transportation easier. Stalks can be immediately chipped on the field by tractor-driven machinery and mobile shredders, making it simpler to load them onto trucks or mini-trolleys [17, 49]. However, transportation is expensive and logistically demanding due to the scattered structure of cotton fields and the low density of unprocessed stalks. In addition to increasing transportation efficiency, briquetting or pelleting lowers feedstock loss while in route [50]. In order to maximize vehicle use and reduce storage obstacles, efficient logistics approaches like modular loading and just-in-time delivery are being investigated. Coordinated collection, better storage infrastructure and transportable processing units can greatly increase the viability of cotton stalk biomass for industrial application, but seasonal supply limitations and the absence of established supply chains continue to be major obstacles. [17, 49, 50].

5 Environmental and Socioeconomic Impacts

5.1 Environmental Benefits

5.1.1 Mitigating GHG Emissions by Avoiding Open-Field Burning

Utilizing agricultural residues for bioenergy generation mitigates significant GHG emissions and various other environmental threats caused by open burning. Open burning of crop residues, particularly cotton stalks, is a major source of GHG emissions and widespread air pollution, releasing substantial quantities of CO_2, methane (CH_4), and nitrous oxide (N_2O) into the atmosphere. By diverting and utilizing these residues into bioenergy pathways, emissions from both fossil fuel replacement and residue burning are avoided, resulting in a net positive climate impact [51–53]. Research shows that effective utilization of agricultural biomass for energy can replace coal and other fossil fuels, leading to significant reductions in CO_2, CH_4, and N_2O emissions [52]. One of the studies found that CH_4 emissions could be reduced by up to 12 million tons and N_2O by 185,000 tons annually if agricultural residues are utilized for bioenergy production instead of being burned in the field. This transition also helps achieve carbon peak and carbon neutrality targets by promoting a low-carbon, circular economy [52, 53]. Sustainable exploitation of agricultural residues for bioenergy has been proven to alleviate nitrous oxide emission levels by 18% and reduce mineral fertilizer demand by 37%, further mitigating indirect GHG emissions from fertilizer production and consumption [54]. Projects like the URJA initiative in India have validated that converting cotton stalks into bioenergy products not only eradicates the practise of open burning (responsible for 27% of residue burning in some regions) but also provides additional environmental benefits such as reduced fertilizer use, increased water efficiency and overall improvement of soil health [8].

5.1.2 Biochar as a Sustainable Approach to Soil Amendment

Biochar is increasingly recognized as a useful and important additive with several advantages for crop yield and soil fertility. It is derived from agricultural residues, particularly cotton stalks and is proven to play a significant role in soil amendment.

Cotton stalk biochar is a stable, nutrient-rich material for agricultural application that is made by pyrolysis, a thermal process that does not use oxygen. This method preserves the carbon-rich structure of the original biomass [55, 56] Cotton stalk biochar improves the physicochemical and biological characteristics of soil and raises its organic carbon content, which increases soil fertility [55, 57, 58]. Its large surface area and high porosity enable water retention, reduce nutrient leaching, and thereby improve nutrient availability, which collectively improve plant development and crop yields [55, 59]. Field trials have shown that biochar from cotton stalks can significantly enhance soil structure, leading to better root development and increased microbial activity, both of which are critical for sustained agricultural productivity [60, 61]. Along with fertility benefits, biochar improves soil quality and reduce environmental threats by acting as an efficient sorbent that captures and immobilizes pollutants (heavy metals, pesticides, and emerging contaminants) [62]. By carbon sequestration (storing carbon in a stable form) and reducing atmospheric CO_2 levels, the application of biochar also help to mitigate climate change and provide farmers with potential carbon credits [55, 59]. Biochar's impact varies depending on its source, particle size, and application rate, as well as inherent soil characteristics and cropping systems [62]. While generally beneficial, excessive application or poor-quality biochar may negatively affect soil organisms or nutrient balance, underscoring the need for site-specific management [59]. Overall, cotton stalk biochar presents a sustainable solution for agricultural waste management, soil enhancement, and climate resilience, with growing evidence supporting its adoption in diverse agroecosystems.

5.1.3 Improved Air and Soil Quality

The air and soil quality is greatly improved when agricultural residues are not burned in the open but are instead utilized for bioenergy generation. Open burning of residues such as cotton stalks releases GHGs, CO, PM and volatile organic compounds (VOCs), all of which contribute to smog formation and resultant respiratory health problems in surrounding communities [63, 64]. Cleaner air and a healthier environment are achieved by drastically reducing harmful emissions, by diverting these agricultural residues to bioenergy production [63, 65, 66]. When combined with site-specific agricultural methods, sustainable management and the use of crop residues for bioenergy can improve both air and soil quality [54]. In comparison to traditional open burning or unorganized residue removal, studies show that integrating residue removal with practices like optimized fertilization, humus balancing and winter soil cover increases soil organic carbon (SOC) by 8% as well as reduces nitrous oxide emissions and nitrate leaching by 18% and 16%, respectively. By reducing nutrient runoff, these methods preserve soil fertility, cut down on the fertilizers demand by up to 37%, and protect water quality [54]. Moreover, digestate or biochar (by-products from bioenergy processes) can be returned to the fields so as to enhance soil nutrient content, water retention and further improving soil structure [67]. This integrated method not only prevents the deterioration associated with residue burning but also promotes long-term agricultural productivity and environmental sustainability. Thus, shifting from open burning to bioenergy utilization of agricultural residues is a proven strategy for achieving cleaner air and healthier, more resilient soils [54, 67].

5.2 *Socioeconomic Benefits*

5.2.1 Income Generation for Farmers

Farmers are becoming more aware that using agricultural waste residues, especially cotton stalks, can be a direct way to generate revenue. Cotton stalks, which were once considered as a waste or burned in the open fields, are now being converted into useful materials for industrial and bioenergy applications, creating new sources of income. Innovative initiatives like as the URJA have shown that the use of mobile pelletizer machines, which enable on-site processing even in remote locations, can help farmers boost their yearly revenue by an average of ₹14,300 through improvement in crop yields and the sale of carbon credits [8]. The market price of cotton stalks ranges from ₹3 to ₹3.5 per kg in raw form and up to ₹7 per kg as briquettes. Cotton stalks can yield at least three tons/ha, with higher yields in irrigated areas [48, 68]. Farmers can earn a minimum of ₹500 per ton when selling unprocessed stalks, while cleaned and chipped stalks transported to mills can fetch up to ₹1200 per ton [68]. In addition to direct sales, cotton stalks are processed into briquettes particle boards, pellets, biochar and bioethanol, encouraging rural entrepreneurship and self-employment [49, 68]. By providing raw material for over 1000 industrial products and renewable energy, farmers benefit from diversified income sources, new employment opportunities and reduced waste, demonstrating the economic potential of agricultural residue utilization, especially cotton stalks [49].

5.2.2 Decentralized Energy and Rural Electrification

In India and other cotton producing regions, the use of agricultural residues (especially cotton stalks) has proven to be revolutionary for decentralized energy generation and rural electrification. Through processes like combustion and gasification, cotton stalks an abundant byproduct of cotton farming (with India producing about 20 million tons annually), can be effectively transformed into bioenergy yielding products like briquettes, pellets and syngas [22, 48]. Cotton stalks are a powerful local energy source for rural and off-grid areas, producing 400–500 kWh of electricity or 4000–5000 MJ of thermal energy per ton [18, 48]. Projects like the URJA initiative have proved the social and economic benefits of this approach by converting cotton stalks into bioenergy, which not only empowers rural populations (including tribal women) by employing them across the entire value chain of bioenergy products but also reduces the levels of open burning and associated air pollution [8]. In areas where the electrical grid is often sporadic or nonexistent, the decentralized nature of cotton stalk collection and processing makes it possible to set up microgrids and small-scale, village-level biomass power plants that provide dependable electricity for small industries, lighting, and irrigation [18, 69]. Decentralized bioenergy from cotton stalks supports rural electrification, encourages local entrepreneurship, decreases reliance on fossil fuels, and promotes sustainable development in agricultural communities, according to research and field experience [69, 70]. This concept promotes inclusive rural growth and provides access to clean energy by recognizing a widely available agricultural byproduct.

6 Case Studies from Cotton-Producing Countries

The diverse and growing importance of agricultural residues, particularly cotton stalks, in the production of bioenergy is demonstrated by case studies from India, China and Pakistan emphasizing technological development, socioeconomic advantages, and environmental benefits. Table 3 presents case studies from India, China and Pakistan that demonstrate the diverse uses of cotton stalk valorization.

6.1 India: Industrial Integration and Rural Empowerment

India, the world's largest cotton producer, produces around 20 million tons of cotton stalks annually, of which 20–25% currently used for biomass energy generation [48, 71]. Cotton stalks are converted into briquettes or pellets, providing fuel for power

Table 3 Case studies on cotton stalk bioenergy utilization in major cotton-producing countries

Country	Key applications and outcomes	References
India	~20 million tons/year of cotton stalks generated, out of which 20–25% utilized for bioenergy.	[8, 49, 71, 72]
	Conversion to briquettes/pellets supplies thermal energy (4000–5000 MJ/ton) and electricity (400–500 kWh/tonne).	
	URJA project: 20% increase in cotton yields, 10% less fertilizer use, ₹14,300 annual income rise for farmers, jobs for 1000+ tribal women.	
	1400+ green jobs and ₹9.2 crore in rural income.	
China	Strong policy support for biomass power plants utilizing cotton stalks.	[73]
	Direct combustion in Beijing-Tianjin-Hebei region equivalent to millions of tons of coal.	
	Government incentives attract investment, though logistics and supply chain constraints prevail.	
	Improves rural sustainability and reduces coal dependency.	
Pakistan	Community-level use of cotton stalks for biogas generation.	[74]
	400 kg of stalks = 50 m^3 biogas/day, enough for 20 families.	
	Generates biofertilizer and supports biodegradable product development.	
	Income benefits for over 1000 farm owners and rural workers.	

generation and industrial heating across various sectors like textiles, sugar, and food processing. One ton of cotton stalks can produce around 400–500 kWh of electricity or 4000–5000 MJ of thermal energy, which is 5–8% less expensive than coal [48]. By transforming cotton stalks into pellets and biochar, the URJA Project in Maharashtra demonstrates a comprehensive strategy to address air pollution, enhance soil health, and empower smallholder farmers and tribal women. In addition to providing average yearly income boost of ₹14,300 to the participating farmers and an additional income for more than 1000 tribal women, this initiative has improved cotton yields by 20%, reduced fertilizer use and water consumption by 10% and 15%, respectively [8]. Large-scale supply chains, such a cotton stalk enterprise that produces 450 tons per day, have produced ₹9.2 crore in rural benefits and over 1400 green jobs [72].

6.2 China: Policy-Driven Biomass Power Expansion

China's strategy leverages robust legislative support to encourage the production of biomass electricity from agricultural residues, like cotton stalks, especially in the major cotton-producing regions of the country. The Beijing-Tianjin-Hebei region possesses crop residue energy potential equal to millions tons of coal, making it a common practice to directly burn densified cotton stalks in the form pellets or briquettes in biomass power plants [73]. Investment in large-scale biomass plants has been incentivized by government subsidies and regulations for the buying of renewable energy. Nonetheless, challenges still remain in ensuring a consistent supply of feedstock, necessitating effective systems for distribution, storage, and collection. According to Chinese case studies, incorporating cotton stalks to the energy mix enhances air quality and promotes rural sustainability by diversifying rural incomes, lowering reliance on coal, and reducing open burning [73].

6.3 Pakistan: Community-Scale Biogas and Value-Added Products

Cotton stalks and other agricultural residues are increasingly recognized as resources for decentralized bioenergy in countries like Pakistan, where there is are acute energy shortages [74]. The use of cotton stalks in biogas plants is demonstrated in case studies, where 400 kg of agricultural waste can yield biogas of 50 m^3/day, that is sufficient to provide 20 households with clean cooking energy. Organic fertilizer is also produced as a by-product. According to Saeed et al. [74], other initiatives concentrate on converting cotton residues (particularly cotton stalks) into biodegradable goods and bioenergy, generating additional sources of income for thousands of rural laborers and farm owners. These initiatives not only solve issues with waste management and energy accessibility, but they also promote rural employment and alleviate the harmful environmental effects of residue burning.

7 Policy Recommendations and Future Directions

7.1 *Need for Farmer Awareness Programs*

Stubble burning is a major contributor to GHG emissions, air pollution as well as soil degradation, and is particularly prevalent in northern India. Due to lack of affordable alternatives, limited awareness of environmental as well as health effects and time constraints between two crop growing seasons, many farmers continue this practice despite its harmful effects [75–77]. Research indicates that most farmers are either not aware of the environmental impacts as well as overall health consequences of stubble burning or lack knowledge about alternative techniques of residue management. Research investigations in Haryana indicated that some farmers despite recognizing the negative impacts of stubble burning, are often not aware of feasible alternatives or hold fallacies about the efficiency of new innovations like Happy Seeder technique [77]. Thus education, communication, robust information knowledge along with "Information, Education, and Communication" (IEC) campaigns are important and necessary. This will not only help the farmers know about the consequences of stubble burning but also demonstrate the agronomic as well as economic benefits of residue management [75, 77]. Official frameworks, developed by the Commission for Air Quality Management (CAQM) and reinforced by the Ministry of Environment, Forests and Climate Change (MoEFCC), prioritize crop residue management as an important strategy for controlling air pollution. These frameworks accentuate the importance of IEC campaigns, capacity-building workshops and on-ground demos to encourage both ex-situ and in-situ management of stubble [78]. Training sessions, mass media and kisan camps are recommended by both government and research organizations to reach farmers and encourage them for adoption of various sustainable practices [75, 76]. Research also makes it evident that awareness programs, combined with policy incentives and support for alternative technologies, make contribution to a considerable reduction in stubble burning occurrences in recent years [76, 78]. However, continued efforts are necessary to eradicate this practice entirely and encourage the productive utilization of agricultural residues for bioenergy generation and various value-added applications [75–77].

7.2 *National and Local Policy Frameworks to Encourage Biomass Use*

National and state-level policy frameworks have a critical role in accelerating the acceptance and implementation of agricultural biomass for bioenergy generation in India. The Ministry of New and Renewable Energy (MNRE), at the national level, has systematized several key initiatives. The National Bioenergy Programme, which has been extended through 2026, offers significant financial investments to support the energy recovery processes using urban and industrial biowaste, cattle dung and agricultural residues. For making bioenergy projects further economically appealing, this program is complemented by targeted fiscal incentives like tax relief on compressed biogas (CBG)-based fuel blending and concessional custom duties on machinery imports [79]. By mandating preferential tariffs and permitting the declaration of biomass co-firing in coal-based power plants as eligible for renewable purchase obligations (RPOs), laws like the Electricity Act of 2003 and the National Tariff Policy of 2016 have established a supportive regulatory environment. With an emphasis on distributed generation and rural electrification, the Waste to Energy Programme and the Biomass-based Cogeneration Programme further encourage as well as incentivize agricultural and industrial waste conversion into biogas, bio-CNG and electricity [79, 80]. Policies at the state level include state-specific preferential tariffs as well as wheeling, buyback and banking arrangements for electricity produced from biomass. Energy service companies (ESCOs), non-governmental organizations (NGOs) and co-operatives frequently assist local governments and agencies, especially village panchayats, in their demonstration projects and decentralized energy initiatives [80]. In order to strengthen the financial support systems, the Indian Renewable Energy Development Agency (IREDA) provides loans for cogeneration and biomass power projects. Despite these frameworks, technological as well as institutional barriers and market challenges still remain. However in addition to supporting India's renewable energy and climate goals these coordinated policy measures have generated more investment, jobs and rural development [80–82].

7.3 *Investment in Research and Development for Advanced Technologies*

Investment in research and development (R&D) is essential for the advancement of technology that enable generation of bioenergy efficiently from agricultural residues. This commitment is demonstrated by India's National Bioenergy Programme (NBP), administered by the Ministry of New and Renewable Energy (MNRE), by supportive R&D initiatives and pilot projects focused on innovative conversion technologies such as biomethanation, pyrolysis, gasification and direct combustion. Central Financial Assistance (CFA) supports these initiatives by boosting the growth and development of these advanced bioenergy solutions in the form of grants and capital subsidies [83, 84]. Recent developments validate the importance of focused R&D investment. Recently, Indian scientists have developed a novel two-step process for direct hydrogen generation from agricultural left-over that escalates efficiency by 25% over conventional anaerobic digestion and prevents the need of expensive pre-treatment [85]. This technology proves several benefits of advanced R&D by generating nutrient-rich digestate applied as an organic fertilizer along with hydrogen production, a fundamental fuel for transition to clean energy [85]. Collaborative projects, like the UK-India Rural Hybrid Energy Enterprise Systems (RHEES), further illuminate the significance of international research collaborations for development of hybrid renewable energy systems combining solar energy with biomass for rural electrification and entrepreneurship [86]. Large-scale adoption and implementation of bioenergy technologies relies on bridging the research gaps related to process optimization, feedstock variability and supply chain integration [86, 87]. Moreover, R&D investment encourages innovation in biochemical and thermochemical conversion methods, which empowers the valorization of several types of residues like cotton stalks, rice straw and sugarcane bagasse [88]. India's regulatory framework ensures that R&D will continue to boost economic feasibility, technological advancement and environmental sustainability in the bioenergy sector by supporting skill development, capacity building and public-private partnerships [87].

7.4 Aligning Bioenergy Initiatives with Textile Sustainability Goals

Aligning bioenergy initiatives with textile industry's sustainability goals provides a synergistic approach in reducing the environmental footprint of the industry while improving resource efficiency. Textile production and processing generates substantial waste, including cotton stalks, fabric scraps and dyeing effluents, which have tremendous valorization potential in generating bioenergy through pyrolysis, gasification or anaerobic digestion. This circular approach primarily diverts waste from dumping into landfills, reduces GHG emissions and alleviates the reliance on fossil fuel, thus directly supporting decarbonization goals. For instance, 30% of natural gas consumption in textile mills can be replaced by converting cotton waste into syngas or biogas in the form of thermal energy in various processes like dyeing and finishing, thus significantly cutting Scope 1 emissions [13, 89]. Incorporating bioenergy into biomass-derived fibers further supports sustainability. Innovations in biomass polyamides and polyesters, synthesized from agricultural residues, reduce dependence on petroleum-based products while continuing to perform on comparable levels with conventional fabrics [90]. Such materials align with circular economy principles, as post-consumer textile waste can be recycled into bioenergy feedstocks, closing the loop. As a soil amendment to increase crop yields and as a carbon sequestrate, biochar produced from cotton stalk pyrolysis aids both the purposes very well, thus mitigating the industry's agricultural carbon footprint [89]. Policy frameworks and technological R&D are critical enablers Textile clusters are encouraged to implement decentralized bioenergy systems by national bioenergy programs, such as India's efforts to encourage biomass co-firing in industrial boilers. These solutions not only promote rural electrification and employment in biomass supply chains but also provide renewable energy in the form of heat and electricity [89]. Advanced biorefineries are emerging to process textile waste into biohydrogen, bioethanol and other biochemicals, thus expanding revenue sources while reducing waste [13, 91].

8 Conclusion

The valorization of cotton stalks signifies a transformative prospect to address prevailing environmental, economic as well as industrial challenges and issues associated with management of agricultural waste. Despite being a major byproduct of cotton cultivation in major cotton-producing nations like India, China, and Pakistan, cotton stalks have long been considered as waste, are frequently incinerated in open fields, which results in release of greenhouse gases, air pollution and degradation of soil. However, technological developments and novel research have confirmed that these wastes have a great potential of being a valuable as well as sustainable source of raw material for bioenergy generation, thus aligning with sustainable development goals (SDGs) as well as circular economy principles which help mitigate climate change. From an environmental viewpoint, conversion of cotton stalks into bioenergy via biochemical pathways such as anaerobic digestion or thermochemical processes of pyrolysis and gasification, along with mitigating significant GHG emission, also help alleviate effects of open-field burning. It is estimated that 1 million ton of cotton stalks when openly burnt releases about 0.85 million metric tons of CO_2e. Hence avoiding this practice can prevent release of these high carbon emissions. Moreover, by-products from cotton stalks such as biochar, can be applied as a soil amendment, which enriches the soil with nutrients and enhances moisture retention. This improves the soil health, along with long-term carbon sequestration. From an economic perspective, the utilization of cotton stalks offers an additional source of income for farmers. Initiatives like the URJA project in India have proved that farmers can earn an extra ₹14,000 per year by selling these cotton stalks and carbon credits. By-products derived from cotton biomass like briquettes, syngas or pellets have facilitated rural electrification, created green jobs throughout value chains as well as caused women empowerment by encouraging women-led startups and microenterprises,. The incorporation of cotton-based bioenergy into textile manufacturing and production clusters has alleviated the dependency on fossil fuels and enhanced energy efficiency. The transition to bioenergy has enormous potential, particularly for textile sector. Because of its high thermal energy requirements, especially during dyeing and finishing processes, applying biomass-derived energy instead of fossil fuels can alleviate GHG emissions and improve sustainability. Furthermore, aligning textile waste valorization with bioenergy initiatives supports the broader objectives of textile industry which include waste minimization, resource efficiency and climate responsibility. A determined and concerted call to action is necessary to fully harness the potential of cotton stalk and its valorization. Expansion of financial incentives, streamlining biomass collection and logistics as well as enforcement of stricter regulations against open burning are all priorities of policymakers. Industry leaders should make investments in scalable bioenergy technologies, incorporate renewable energy into production processes, and encourage local-regional supply chains. It is recommended that researchers develop effective pre-treatment and conversion technologies, evaluate lifetime impacts and guide evidence-based policy. Cotton stalks can be transformed from an environmental burden into a sustainable asset by combining the government efforts, academic institutions and industries. This will promote rural livelihoods, encourage clean energy and meaningfully contribute to global sustainable development and climate-resilient goals.

References

1. Ramanjaneyulu AV, Ramprasad B, Sainath N, Umarani E, Pallavi C, Vijay J, Jagadeeshwar R (2021) Crop residue management in cotton. Chron Bioresour Manage 5(1):1–8
2. Silva JE, Deus Junior JO, Calixto GQ, Melo DM, Melo MA, Júnior VC, Braga RM (2024) Colored cotton crop wastes valorization through pyrolysis: a study of energetic characterization and analytical Py-GC/MS. Sci Rep 14(1):9359
3. Velmourougane K, Blaise D, Savitha S, Waghmare VN (2021) Valorization of cotton wastes for agricultural and industrial applications: present status and future prospects. In: Bhat R (ed) Valorization of agri-food wastes and by-products. Academic, pp 665–692. https://doi.org/10.1016/B978-0-12-824044-1.00043-X
4. Babou C, Gururajan B, Suresh P (2005) Effect of cotton crop-residue management practices and levels of nitrogen and potassium on rice (*Oryza sativa*) and succeeding blackgram (*Phaseolus mungo*) in rice–blackgram cropping system. Indian J Agron 50(4):260–264
5. Vitale GS, Scavo A, Zingale S, Tuttolomondo T, Santonoceto C, Pandino G, Guarnaccia P (2024) Agronomic strategies for sustainable cotton production: a systematic literature review. Agriculture 14(9):1597
6. Bhuvaneshwari S, Hettiarachchi H, Meegoda JN (2019) Crop residue burning in India: policy challenges and potential solutions. Int J Environ Res Public Health 16(5):832
7. Tandon SK, Sundaramoorthy C (2018) Environmental preservation through use of cotton stalks for industrial purpose. International Workshop on Utilization of Cotton Plant By-Produce for Value-Added Products, International Cotton Advisory Committee (ICAC). https://www.icac.org/projects/CommonFund/20_ucbvp/papers/09_tandon.pdf
8. Daga MP (2024) URJA Project: transforming cotton stalk waste into bioenergy, biochar, and carbon credits. Cotton Statistics & News 25
9. Cai C, Wang Z, Ma L, Xu Z, Yu J, Li F (2024) Cotton stalk valorization towards bio-based materials, chemicals, and biofuels: a review. Renew Sust Energ Rev 202:114651
10. Cho EJ, Lee YG, Song Y, Kim HY, Nguyen DT, Bae HJ (2023) Converting textile waste into value-added chemicals: an integrated bio-refinery process. Environ Sci Ecotechnol 15:100238
11. Yousef S, Eimontas J, Striūgas N, Tatariants M, Abdelnaby MA, Tuckute S, Kliucininkas L (2019) A sustainable bioenergy conversion strategy for textile waste with self-catalysts using mini-pyrolysis plant. Energy Convers Manage 196:688–704
12. Tharmarajah N, Shahbaz K, Baroutian S (2024) Textile waste pretreatment for anaerobic digestion: a review and technology feasibility study. J Chem Technol Biotechnol 99(11):2213–2227
13. Khandaker S, Bashar MM, Islam A, Hossain MT, Teo SH, Awual MR (2022) Sustainable energy generation from textile biowaste and its challenges: a comprehensive review. Renew Sust Energ Rev 157:112051
14. Li G, Wang Y, Yu D, Zhu P, Zhao G, Liu C, Zhao H (2022) Ligninolytic characteristics of Pleurotus ostreatus cultivated in cotton stalk media. Front Microbiol 13:1035040

15. Zhong P, Chen P, Huo P, Ma L, Xu Z, Li F, Cai C (2025) Characterization of cotton stalk as a lignocellulosic feedstock for single-cell protein production. Bioresour Technol 417:131797
16. Hou X, Sun F, Zhang L, Luo J, Lu D, Yang Y (2014) Chemical-free extraction of cotton stalk bark fibers by steam flash explosion. BioRes 9(4):6950–6967
17. Umesh D, Sarsavadiya P, Vaja K, Mahadeo K (2015) Physiochemical properties of cotton stalk biomass from aricultural residues. Curr World Environ 10(1):343–349
18. Raju NP, Raju M (2019) Experimental study of cotton stalk pellet renewable energy potential from agricultural residue woody biomass as an alternate fuel for fossil fuels to internal combustion engines. Int J Eng Res Technol 8:746–750
19. Gravalos I, Xyradakis P, Kateris D, Gialamas T, Bartzialis D, Giannoulis K (2016) An experimental determination of gross calorific value of different agroforestry species and bio-based industry residues. Nat Resour 7(1):57–68
20. Makavana JM, Agravat VV, Balas PR, Makwana PJ, Vyas VG (2018) Engineering properties of various agricultural residue. Int J Curr Microbiol App Sci 7(6):2362–2367
21. Sakariya KK, Sarsavadia PN, Chauhan PM, Kelaiya SV, Ramani ML (2019) Physicochemical properties and proximate analysis of cotton stalks biomass. Int J Chem Stud 7(6):1375–1378
22. Elshareef H (2022) Bioenergy potential from cotton stalks gasification. In: 14th International conference on biofuels and bioenergy. [Webinar]. Conference series. https://biofuels-bioenergy.conferenceseries.com/
23. Khalil A, Bakry AI, El-Fakharany MK, Rabea K (2018) Experimental study of cotton stalks gasification in a downdraft reactor. J Eng Res 2:1–15
24. Vaskalis I, Zabaniotou A (2024) Comparative feasibility and environmental life cycle assessment of cotton stalks gasification and pyrolysis. Biomass 4(1):23–48
25. Al Afif R, Anayah SS, Pfeifer C (2020) Batch pyrolysis of cotton stalks for evaluation of biochar energy potential. Renew Energy 147:2250–2258
26. Kataria G, Sharma A, Joshi JB, Hameed S, Amiri A (2022) A system level analysis of pyrolysis of cotton stalk biomass. Mater Today Proc 57:1528–1532
27. Kantarelis E, Zabaniotou A (2009) Valorization of cotton stalks by fast pyrolysis and fixed bed air gasification for syngas production as precursor of second generation biofuels and sustainable agriculture. Bioresour Technol 100(2):942–947
28. Al Afif R, Pfeifer C, Pröll T (2019) Bioenergy recovery from cotton stalk. IntechOpen. https://doi.org/10.5772/intechopen.88005
29. Akhtar A, Krepl V, Ivanova T (2018) A combined overview of combustion, pyrolysis, and gasification of biomass. Energy Fuel 32(7):7294–7318
30. Perera SM, Wickramasinghe C, Samarasiri BKT, Narayana M (2021) Modeling of thermochemical conversion of waste biomass–a comprehensive review. Biofuel Res J 8(4):1481–1528
31. Danso-Boateng E, Achaw OW (2022) Bioenergy and biofuel production from biomass using thermochemical conversions technologies—a review. AIMS Energy 10(4):585–647
32. Adekunle KF, Okolie JA (2015) A review of biochemical process of anaerobic digestion. Adv Biosci Biotechnol 6(03):205
33. Archana K, Visckram AS, Kumar PS, Manikandan S, Saravanan A, Natrayan L (2024) A review on recent technological breakthroughs in anaerobic digestion of organic biowaste for biogas generation: challenges towards sustainable development goals. Fuel 358:130298
34. Van DP, Fujiwara T, Tho BL, Toan PPS, Minh GH (2020) A review of anaerobic digestion systems for biodegradable waste: configurations, operating parameters, and current trends. Environ Eng Res 25(1):1–17
35. Monlau F, Sambusiti C, Antoniou N, Barakat A, Zabaniotou A (2015) A new concept for enhancing energy recovery from agricultural residues by coupling anaerobic digestion and pyrolysis process. Appl Energy 148:32–38
36. Bušić A, Marđetko N, Kundas S, Morzak G, Belskaya H, Ivančić Šantek M, Šantek B (2018) Bioethanol production from renewable raw materials and its separation and purification: a review. Food Technol Biotechnol 56(3):289–311

37. Devarapalli M, Atiyeh HK (2015) A review of conversion processes for bioethanol production with a focus on syngas fermentation. Biofuel Res J 2(3):268–280
38. Jaiswal K, Girdher H, Sahu A, Yadav A, Pali HS (2017) Review paper on production of bioethanol. Int J Adv Res Sci Eng 6(2):1–10
39. Amrutrao A, Degnet DD, Gota GY (2020) Use of energy in textile industry: a case study on Ethiopian textile industry. IOSR-JMCE 17(3, Ser. I):58–64. https://www.iosrjournals.org/iosr-jmce/papers/vol17-issue3/Series-1/I1703015864.pdf
40. Kousar S, Shafqat U, Kausar N, Pamucar D, Karaca Y, Salman MA (2022) Sustainable energy consumption model for textile industry using fully intuitionistic fuzzy optimization approach. Comput Intell Neurosci 2022(1):5724825
41. Uddin F (2014) Energy management and energy crisis in textile finishing. Am J Energy Res 2(3):53–59
42. Kasturi S, Kannappan BA (2020) Energy diagnosis and savings potential in knitted fabric based wet processing units. Energy Sustain Dev 54:85–100
43. Bhuiya KMS, Azad AMAS, Arafat I, Hasan M, Udoy SA, Diganta AC, Rana MS, Towfiq MHI, Islam MA, Tamal OH, Masuk NI, Limon MKH (2024) Assessment of energy consumption and emissions of fabric production lines in Bangladesh. In: Proceedings of the 14th International conference on mechanical engineering (ICME 2023). SSRN
44. Prasanga WC, Nawarathna A, Rathnasiri P (2021) Use of cotton apparel waste as an energy source for biomass boilers: a feasibility study. IJDNE 16(1):41–51
45. Kramens J, Vigants E, Kanukuntla SP, Goljandin D (2023) Efficient energy recovery from textile waste and biomass mixture. In: Proceedings of the 22nd International scientific conference "Engineering for rural development". Jelgava, Latvia, pp 24–26
46. Thermax Limited (2024) Navigating energy transition: comprehensive solutions for textile players in the pursuit of sustainability. https://www.thermaxglobal.com/navigating-energy-transition-comprehensive-solutions-for-textile-players-in-the-pursuit-of-sustainability/
47. Hasanbeigi A, Springer C, Wei D (2024) Low-carbon thermal energy technologies for the textile industry. Global Efficiency Intelligence, Commissioned by the Apparel Impact Institute. https://www.globalefficiencyintel.com/lowcarbon-thermal-energy-technologies-for-the-textile-industry
48. Narsi (2024) Cotton stalk in India—availability, supply chain, prices, surplus. In: Biomass Feedstock in India. BioBiz. https://biobiz.in/s/bring/in/9
49. Saxena S (2024) Cotton plant stalks: a valuable industrial raw material. Cotton Statistics & News 11
50. Ravula PP, Grisso RD, Cundiff JS (2008) Cotton logistics as a model for a biomass transportation system. Biomass Bioenergy 32(4):314–325
51. Gowen D (2017) Mobilization of agricultural residues for bioenergy and bioproducts. IEA Bioenergy. https://coilink.org/20.500.12592/3rr1qn
52. Li D (2022) Analysis of agricultural biomass energy use and greenhouse gas reduction evidence from China. J Environ Public Health 2022(1):6126944
53. Zhang J, Li YE, Cai A, Oosterveer P, Greene M, Wang B (2023) Greenhouse gas reduction through crop residue-based bioenergy: a meta-analysis of reduction efficiency and abatement costs of various products. Energy 270:126900
54. Mouratiadou I, Stella T, Gaiser T, Wicke B, Nendel C, Ewert F, van der Hilst F (2020) Sustainable intensification of crop residue exploitation for bioenergy: opportunities and challenges. GCB Bioenergy 12(1):71–89
55. Pawar D (2024) Harnessing the power of cotton stalk biomass: EcoKrishi's biochar. EcoKrishi Solutions. https://www.ecokrishi.in/post/unlocking-the-potential-of-biochar-transforming-agriculture-with-ecokriahi
56. Phadtare P, Kalbande S (2024) Assessment of agro-residues as biochar feedstock: cotton stalk, soybean straw, and pigeon pea stalk. Int J Res Agron SP-7(6):36–40
57. Kavitha B, Reddy PVL, Kim B, Lee SS, Pandey SK, Kim KH (2018) Benefits and limitations of biochar amendment in agricultural soils: a review. J Environ Manage 227:146–154

58. Patel MR, Panwar NL (2023) Biochar from agricultural crop residues: environmental, production, and life cycle assessment overview. Resour Cons Recy Adv 19:200173
59. Lamichhane B, Dunn B, Ouedraogo A, Singh H (2023) February. Oklahoma State University, Preparation of biochar for use as a soil amendment. https://extension.okstate.edu/fact-sheets/preparation-of-biochar-for-use-as-a-soil-amendment.html
60. Kabir E, Kim KH, Kwon EE (2023) Biochar as a tool for the improvement of soil and environment. Front Environ Sci 11:1324533
61. Solidaridad Network (2024) Biochar revitalizes cotton farming in Zambia. https://www.solidaridadnetwork.org/news/biochar-revitalizes-cotton-farming-in-zambia/
62. Nepal J, Ahmad W, Munsif F, Khan A, Zou Z (2023) Advances and prospects of biochar in improving soil fertility, biochemical quality, and environmental applications. Front Environ Sci 11:1114752
63. Kumar JA, Sathish S, Prabu D, Renita AA, Saravanan A, Deivayanai VC, Hosseini-Bandegharaei A (2023) Agricultural waste biomass for sustainable bioenergy production: feedstock, characterization and pre-treatment methodologies. Chemosphere 331:138680
64. Prasad S, Singh A, Korres NE, Rathore D, Sevda S, Pant D (2020) Sustainable utilization of crop residues for energy generation: a life cycle assessment (LCA) perspective. Bioresour Technol 303:122964
65. Fitri H, Gürdil GA, Demirel B, Cevher EY, Roubík H (2023) Biomass potential from agricultural residues for energy utilization in West Nusa Tenggara (WNT), Indonesia. GCB Bioenergy 15(11):1405–1414
66. Omer AM (2019) Economic importance of environmental development of bioenergy from agriculture residues and environment. Int J Res Stud Microbiol Biotechnol 5(3):17–42
67. IEA (International Energy Agency) (2005) Benefits of bioenergy bioenergy (ExCo: 2005:01). IEA Bioenergy
68. PTI (Press Trust of India) (2013) Cotton stalks throw open new income stream for farmers. Business Standard. https://www.business-standard.com/article/markets/cotton-stalks-throw-open-new-income-stream-for-farmers-105120201017_1.html
69. FAO (Food and Agriculture Organization of the United Nations) (2019) Off-grid rural electrification options using crop and woody residues in Côte d'Ivoire: Bioenergy and Food Security (BEFS) case study. FAO. https://www.fao.org/energy/news/news-details/en/c/1187366/
70. Tatsiopoulos IP, Tolis AJ (2003) Economic aspects of the cotton-stalk biomass logistics and comparison of supply chain methods. Biomass Bioenergy 24(3):199–214
71. Udakwar S, Sarode D (2023) A study of generation and regulation of crop residue: bioenergy. Bhartiya Krishi Anusandhan Patrika 38(3):267–275. https://doi.org/10.18805/bkap648
72. TERI (The Energy and Resources Institute) (2017) Workshop on effective and efficient realization of the extensive bioenergy resources of India: proceedings. TERI
73. Negi H, Suyal DC, Soni R, Giri K, Goel R (2023) Indian scenario of biomass availability and its bioenergy-conversion potential. Energies 16(15):5805
74. Saeed MA, Irshad A, Sattar H, Andrews GE, Phylaktou HN, Gibbs BM (2015) Agricultural waste biomass energy potential in Pakistan. In: Proceedings of the international conference. University of Leeds, Shanghai
75. Kadian M, Nagoria S, Monga S (2024) Stubble burning in India: environmental concern and alternative tools. Curr Agric Res J 12(1):161–169
76. Kohli R, Mittal A, Mittal A (2024) Research insights into Punjab's stubble burning menace. Nat Environ Pollut Technol 23(3):1687–1696
77. Sharma A (2020) Stubble burning: the need for a behavioural lens. Social and Political Research Foundation. https://sprf.in/stubble-burning-the-need-for-a-behavioural-lens/
78. MoEFCC (Ministry of Environment, Forest and Climate Change) (2024) Air pollution caused by stubble burning. Press Information Bureau, Government of India. https://pib.gov.in/PressReleasePage.aspx?PRID=2035034

79. Pelkmans L (ed) (2021) Implementation of bioenergy in India. Country reports. IEA Bioenergy. https://www.ieabioenergy.com/wp-content/uploads/2021/11/CountryReport2021_India_final.pdf
80. Ravindranath D, Rao SSN (2011) Bioenergy in India: barriers and policy options. In: Haselip J, Nygaard I, Hansen U, Ackom E (eds) Diffusion of renewable energy technologies: Case studies of enabling frameworks in developing countries. UNEP Risø Centre on Energy, Climate and Sustainable Development, Roskilde, pp 129–158
81. Kothari R, Vashishtha A, Singh HM, Pathak VV, Tyagi VV, Yadav BC, Singh DP (2020) Assessment of Indian bioenergy policy for sustainable environment and its impact for rural India: strategic implementation and challenges. Environ Technol Innov 20:101078
82. Shoeb S, Shoeb S, Munwaroddin M, Quadri SS (2017) A review of biomass energy: Indian scenario. Int J Res Eng Technol 6(9):24–28
83. MNRE (Ministry of New and Renewable Energy) (2021) Bio energy overview. Government of India. https://mnre.gov.in/en/bio-energy-overview/
84. MNRE (Ministry of New and Renewable Energy) (2022) National Bioenergy Programme 2022. Government of India. https://mnre.gov.in/en/bio-energy-overview/
85. DST (Department of Science & Technology) (2021) Unique technology for direct generation of hydrogen from agricultural residue developed. Government of India. https://dst.gov.in/unique-technology-direct-generation-hydrogen-agricultural-residue-developed
86. Thornley P, Rathi B, Borah AJ, Saha B, Adams J, Hiloidhari M, Blanchard R, Dinsdale R, Nagarajan S, Kumar S, Vaidyanathan S, Sharma S, Sahota S, Masakapalli SK (2022) Bioenergy technologies for a net zero transition: outcomes of UK-India bioenergy research scoping. Supergen Bioenergy Hub. https://www.supergen-bioenergy.net/wp-content/uploads/2019/06/Bioenergy-Technologies-for-a-Net-Zero-Transition-UK-India-Final-Report-2022.pdf
87. Narsi (2024) India agricultural residues management: market potential, costs, companies, technology. Bio Environmental Solutions. https://biobiz.in/s/opp/agricultural-residues-management
88. Sivatmiga TN, Bindu B, Kumar NR, Srimathi P, Mullainathan VH (2023) Bioenergy generation from agricultural residues: a comprehensive review of recent advances in India. Int J Novel Res Dev 8(10):1–9
89. Nunes LJR, Godina R, de Oliveira Matias JC (2019) Technological innovation in biomass energy for the sustainable growth of textile industry. Sustainability 11(2):528
90. Fibre2Fashion (2013) Bio-mass fibres in textiles. https://www.fibre2fashion.com/industry-article/7045/bio-mass-fibres-in-textiles
91. Novonesis (2024) How biology can make textile production more sustainable. The Biosolutions Bulletin. https://www.novonesis.com/en/news/the-biosolutions-bulletin-07-how-biology-can-make-textile-production-more-sustainable

Renewable Energy in Textiles: Addressing Climate Change

Samar Gogari Radha

Abstract The textile sector is a major contributor to global greenhouse gas emissions because of its high dependence on fossil fuels for energy-intensive processes like dyeing, spinning, and finishing. In the wake of climate change, incorporating renewable energy sources into textile manufacturing offers a sustainable and economically feasible solution. This chapter explores the environmental footprint of conventional textile energy sources and highlights the shift towards solar, wind, geothermal, and biomass energy in textile production. The implementation of renewable energy along with reducing the carbon footprint also enhances operational efficiency and enduring productivity. Technological developments, like energy-efficient machinery and green chemistry, are discussed in the context of energy demand reduction. The chapter also studies international and national policies that support sustainable energy in the textile industry. Furthermore, it evaluates the socio-economic advantages like employment, health enhancements and community upliftment in high textile active regions. Despite its potential, the transition has obstacles like high capital costs, technical barriers, and policy gaps. The chapter concludes by recommending integrated renewable energy strategies, supported by regulatory frameworks and technological innovation, to accelerate the textile industry's contribution to climate mitigation.

Keywords Climate change · Greenhouse gases · Textile industry · Renewable energy · Carbon footprint

S. G. Radha (✉)
School of Biological and Environmental Sciences, Shoolini University of Biotechnology and Management Sciences, Solan, Himachal Pradesh, India

S. S. Muthu (ed.), *Renewable Energy Implications for the Textile Industry*, Sustainable Textiles: Production, Processing, Manufacturing & Chemistry,
https://doi.org/10.1007/978-3-032-10917-0_5

1 Introduction

1.1 Global Energy Consumption in the Textile Industry

The textile industry is one of the major energy consumer in the world, with its energy demand influenced by the scale and diversity of its operations. The industry's enormous facilities and continuous expansion causes significant levels of energy consumption, but the energy intensity per unit product is quite lower than other heavy industries. Globally, the textile industry accounts for about less than 2% of final energy use for production in the U.S. but around 4% in China, thus reflecting differences in production scale and industrial structure [1]. Energy expenditures typically comprises of 15–20% of total manufacturing costs in various operations [2]. Textile manufacturing is highly dependent on both thermal energy and electricity. Thermal energy is mainly used in the form of steam which is essential for dyeing, wet processing and finishing, while electricity highly utilized in spinning and weaving processes [1–3]. Approximately, 61% of the final energy used in the textile industry globally, comes from fuel and 39% from electricity. Steam-driven systems and motors both combined, nearly account 29% for total final energy [1]. The industry is also one of the major consumers of steam among industrial sectors. The ever-increasing, expansion of fast fashion trend and synthetic fiber production has additionally intensified energy demand and associated greenhouse gas (GHG) emissions. The textile industry contributes nearly 1.2 billion tons of CO_2 equivalent (CO_2e) annually, which is approximately 10% of global GHG emissions [4, 5]. As textiles and apparels demand rises, integrating renewable energy sources and increasing energy efficiency are vital for reducing the sector's environmental footprint and ensuring its long-term sustainability [6].

This chapter examines how renewable energy can reduce the environmental footprint of the textile industry. It covers the impacts of fossil fuel use, the potential of solar, wind, biomass, and geothermal energy, and practical strategies for implementation. Key policies, socio-economic benefits, challenges, and scaling approaches are discussed to support a sustainable and climate-resilient textile sector.

1.2 Role of Renewable Energy in Reducing the Carbon Footprint in Textiles

Majority of the textile industry's carbon footprint comes from raw material production, specifically synthetic fibers derivatives of petroleum and conventional cotton, which collectively represent more than 40% of a garment's total emissions [7, 8]. Manufacturing processes such as spinning, dyeing, and finishing being energy-intensive, are highly dependent on steam and electricity both of which are fossil fuel-based, further escalating the industry's environmental impact [7, 9].

Transitioning to renewable energy in textile manufacturing has a significant potential for reducing the industry's carbon footprint. Studies show that combining renewable electricity and low-carbon thermal energy technologies (solar thermal, biomass, and electrification) can reduce emissions from process heating and electricity use which cause major emissions [8, 10]. Textile companies that have implemented renewable energy sources have reported noteworthy reductions in their total carbon emissions, with some achieving reserves of thousands of tons of CO_2e annually [10]. However, the efficiency of renewable energy implementation depends on factors like the availability of renewable resources, carbon intensity of local electricity grids and the extent of implementation [8]. Widespread implementation of renewables, combined with sustainable material selection and energy efficiency measures, will enable textile sector to reduce its environmental carbon footprint and meet global climate targets [10].

1.3 Importance of Transitioning to Renewable Energy

Immediate necessity to manage operational expenses, reduce carbon emissions and promise long-term sustainability, drive the textile industry's transition to renewable energy. The textile industry consumes large amount of energy due to its energy-intensive processes like spinning, weaving, dyeing, and finishing accounting for a large carbon footprint [11]. In fact, more than 80% of carbon emissions for fashion and apparel brands originate from their supply chains, rendering renewable energy adoption vital for serious and effective climate action [11, 12]. Shifting to renewables such as solar, wind, and residual biomass provides various benefits. Primarily, it causes direct reduction reliance on fossil fuels dependence, alleviating GHG emissions and enabling the industry meet international climate goals as well as commitments [11, 13, 14]. Secondly, renewable energy offers better price stability in comparison with volatile fossil fuel, ensuring healthier financial planning and risk management for companies [12]. Lastly, energy expenditures constitute up to 25% of operational costs in textile manufacturing, and studies show that renewable energy implementation can yield large savings and improve affordability [11, 12]. Furthermore, as global regulations become stricter for emissions and sustainability, early implementation of renewables positions textile industries for compliance, avoiding restrictions and future penalties [12, 13]. Decentralized renewable energy solutions, like rooftop solar and microgrids, also boost energy accessibility and resilience, especially for small-scale and rural textile initiatives [11, 15]. Hence, the transition to renewable energy is both an environmental imperative and a strategic business decision for the textile industry's future.

1.4 Study Objectives and Implication

The primary objective of the study is to thoroughly examine the environmental impacts of the textile industry due to their reliance on fossil fuel and thus explore the importance of renewable energy in climate change mitigation. The study aims to recommend renewable alternatives like solar, wind, biomass, and geothermal technologies by identifying main energy-intensive processes like spinning, dyeing, and finishing within textile manufacturing. A central focus is the combined advantage of renewable energy which is environmental sustainability via carbon emission reduction and economic sustainability through long-term energy and cost efficiency. The study has significant implications for environmentalists, legislators and industry stakeholders. It demonstrates that incorporating renewable energy in the textile industry not only helps in significant carbon footprint reduction but also improves operational resilience and profitability. In textile-intense regions like India, China, and Bangladesh, the study evaluates the socio-economic advantages, like job creation, improvement of overall community health, and energy accessibility. Moreover, the research emphasizes the importance of regulatory frameworks, technological innovation and capacity-building programmes to overcome barriers like policy fragmentation, capital costs and technical limitations. Ultimately, the study necessitates a holistic, multi-stakeholder approach to accelerate the transition of textile sector toward a low-carbon future.

2 Environmental Impact of Conventional Energy Use in Textiles

2.1 Energy Consumption in Critical Textile Processes

Energy consumption in the textile industry is dispersed across numerous important processes, each with distinctive energy profiles and efficiency challenges. Spinning, the most electricity-intensive process, which converts raw fibers into yarn, accounts for around 34–41% of total energy consumption in textile manufacturing [3, 6]. This process is highly dependent on electric motors for driving spinning frames, carding, and drawing machines. Hence, optimization of motor efficiency and reduction of idle periods are critical for saving energy in spinning operations [1, 3]. Weaving, which interweaves yarns for producing fabric, is responsible for 23% of the industry's energy consumption, mainly in the form of electricity for powering looms and supplementary equipment [3, 6]. The energy consumption in weaving depends upon the type of loom, automation level, and manufacturing speed. Upgrading to energy-efficient looms and employing real-time monitoring can lead to considerable decline in electricity consumption [3]. Dyeing and finishing are also high energy-intensive processes, jointly responsible for more than 38% of total textile energy consumption [1, 3, 16]. These processes demand significant thermal energy, primarily steam, for the purpose of heating dye baths, drying, washing, and curing. Dyeing machines like winch, jet, and beam dyeing units need around 1.5–17 GJ/ton of fabric, depending on the temperature and method employed [16, 17]. Finishing processes, like heat-setting and stentering, also require significant heat input for drying and to set fabric properties [17]. Generally, energy expenses represent 15–20% of textile manufacture expenditures, necessitating efficient energy management for competitiveness and sustainability in the textile sector [3, 18].

2.2 Emissions and Wastewater Pollution

The textile industry is a major contributor to global emissions, contributing majorly to the release of CO_2, NO_x, SO_x, and wastewater pollutants. CO_2 emissions predominantly arise from energy-intensive processes like dyeing, finishing, and synthetic fiber production. A study by Quantis stressed that the apparel sector is solely responsible for generating 6.7% of global GHG emissions ($\approx$3.3 billion metric tons CO_2e annually), with dyeing and finishing contributing over 50% of these emissions [19]. Coal-dependent energy systems in major textile-producing nations like China and India exacerbate this footprint, with wet processing (steam heating) accounting for 38% of total energy utilization [5]. NO_x and SO_x emissions are associated with fossil fuel combustion in boilers and chemical reactions during dyeing. Research applying the USEtox model in textile sector of China showed that NO_x emissions, posed a chemical footprint (ChF) eight times greater than SO_2, with the textile manufacturing sub-sector contributing 73% of total ChF [20]. These pollutants are responsible for acid rain, particulate matter (PM) formation, and respiratory diseases. Wastewater from textile processes contains toxically high levels of organic pollutants, dyes, and other toxic pollutants. In a study it was found COD (4449 mg/L), BOD (1840 mg/L), and TDS (21,989 mg/L) in untreated effluent, were exceedingly higher than the safe limits [21]. Such contamination risks aquatic ecosystems and public health as well as sanitation, emphasizing the need for advanced treatment solutions.

2.3 Fossil Fuel Dependency in Textile-Dense Regions

For thermal energy and electricity generation in particular, textile industries in major textile-producing regions like India, China, and Bangladesh are largely dependent on fossil fuels. In Bangladesh, the textile and garment sector predominantly depends on natural gas-based onsite production, which accounts for 90–97% of total energy consumption. Only a small portion of this energy share comes from the diesel, national grid or renewable sources. This dependency on fossil fuels, particularly coal and natural gas, is a major factor for GHG emissions, contributing to both local air pollution and global climate change [22–24]. In China, the expansion of the textile industry has been identified as a predominant factor for increased CO_2 emissions, with the sector's energy demands mostly met by petroleum products and coal. Similarly, Indian textile industries are characterized by extensive use of hard coal and natural gas for heating and electricity purposes, which significantly intensifies the carbon footprint of textile products manufactured in those particular regions [9, 23, 24]. Studies indicated that coal is particularly dominant in the top textile-producing nations, like India, China, and Vietnam, and this is responsible for harmful onsite air pollution and high carbon intensity [9, 23]. This heavy reliance on fossil fuel intensifies the environmental impact of textile production and also exposes the industry to energy price fluctuations and regulatory complications as global climate policies become rigid. Instant transition to renewable energy sources, like solar, wind and biomass, is acknowledged as a key strategy for emission reduction and sustainability improvement in these textile-dense regions [9, 22, 24].

2.4 Case Examples of Unsustainable Practices

Case studies from textile-dense regions reveal several unsustainable practices in energy consumption that majorly impact both operational efficiency and environmental performance. In an Ethiopian textile plant, research showed that thermal energy (primarily fossil fuels-derived), is severely used for chemical processing, predominantly in heating and drying processes. The plant depended on steam provided via inefficient distribution systems, causing significant energy losses and higher consumption rates in wet processing. Additionally, electrical energy use in spinning and weaving was found to be consistently high, with outdated machinery intensifying inefficiencies and increasing overall electricity demand [3]. A Turkish study demonstrated similar trends, where energy consumption and expenses were closely adhering to production levels, but unnecessary waste was generated because of poor insulation and inefficient steam production during the processes of dyeing and finishing [25]. In India, many textile facilities continue to depend on coal-fired boilers for process heating, a practice that not only escalates CO_2 emissions but also results in significant onsite air pollution and energy losses through outdated steam systems [26]. Energy audits in textile mills have frequently recognized issues such as excessive heat loss, deficiency of thermal insulation, prolonged machine idle times, and the use of low-efficiency motors as major contributors to unsustainable energy use [27].

3 Renewable Energy Technologies in Textile Manufacturing

3.1 Solar Energy

Solar energy-based technologies are rapidly transforming the textile manufacturing sector, offering a pathway to decarbonization and operational cost savings. The textile industry's high demand for both electricity and low-to-medium temperature heat makes it ideally suited for solar integration. Two principal solar technologies-solar photovoltaic (PV) and solar thermal-are being widely adopted in textile hubs globally. Solar PV systems supply renewable electricity for spinning, weaving, and finishing operations. Large-scale rooftop and ground-mounted solar installations have been implemented by leading textile companies in India and China, such as Arvind Limited and H&M's suppliers, resulting in annual carbon emission reductions of up to 20,000 metric tons per facility and substantial energy cost savings [28, 29]. In Tamil Nadu, India, textile manufacturers have installed over 100 MW of solar PV capacity, with ongoing projects aiming for even greater uptake [28]. The decreasing cost of solar PV—down by over 80% since 2010—has made this technology highly accessible for both large and small enterprises [29]. Solar thermal systems, including concentrating solar thermal (CST) collectors, provide process heat for dyeing, bleaching, washing, and drying. Research and industry reports indicate that up to 50% of a typical textile mill's thermal energy demand can be met using solar thermal solutions, significantly reducing fossil fuel consumption and GHG emissions [30, 31]. Heat storage technologies further enable continuous operation, overcoming the intermittency of solar supply [30]. Despite challenges such as high upfront investment and integration with existing infrastructure, solar energy adoption in textiles is supported by favourable policies and incentives in key markets. As technology advances, solar solutions are poised to play a central role in the textile industry's transition to sustainability and net-zero emissions [29, 31, 32]. Table 1 gives an overview of renewable energy technologies applied across various processes in textile manufacturing.

3.2 *Wind Energy*

Wind energy-based technologies are increasingly significant in textile manufacturing, both as a source of renewable power for factory operations and as a key area of material innovation within the wind energy supply chain. Textile production is highly electricity-intensive, especially in spinning and weaving processes, making the integration of wind-generated electricity an effective strategy for reducing reliance on fossil fuels and lowering carbon emissions [30]. Globally, the textile industry is not only a major consumer of wind-generated electricity but also a contributor to the wind energy sector through advanced textile materials used in turbine components. Carbon and glass fibers, for example, are essential in manufacturing lightweight, high-strength wind turbine blades. Recent advancements include the use of recycled carbon fibers and natural fiber composites (such as flax), which further reduce the environmental impact of both the textile and wind energy industries. Innovations like the Green Nacelle project demonstrate how bio-based textile composites can lower CO_2 emissions by up to 60% and reduce energy consumption by over 50% compared to conventional materials [33, 34]. The manufacturing of wind turbine components now involves a complex supply chain with over 500 specialized

Table 1 Renewable energy technologies in textile manufacturing

Technology	Applications in textile manufacturing	Advantages	Challenges	References
Solar energy (PV and thermal)	• Solar PV: Powers spinning, weaving, finishing operations • Solar thermal: Supplies heat for dyeing, bleaching, washing, and drying	• Reduces CO_2 emissions by up to 20,000 metric tons per facility • Cuts energy costs • Up to 50% of thermal energy needs can be met • Favorable policies and incentives	• High upfront investment • Integration with existing systems • Intermittent supply (mitigated by heat storage)	[28–32]
Wind energy	• Provides renewable electricity for spinning and weaving • Enables production of eco-friendly composites for wind turbine components	• Reduces reliance on fossil fuels • Supports sustainable materials development • Textile composites reduce turbine blade weight and CO_2 emissions by up to 60%	• Complex supply chains • Requires large-scale infrastructure and consistent wind resource	[30, 33–35]
Biomass energy	• Biomass boilers for process heat and steam generation using agricultural/ textile residues	• Reduces steam costs • Utilizes textile waste • Cuts reliance on firewood • Reduces environmental impact of waste	• Sustainability depends on biomass supply chain • Risk of ecosystem degradation if mismanaged	[9, 36–38]
Geothermal and emerging sources	• Geothermal: Continuous heat for dyeing, drying, washing • Emerging: Electric boilers, industrial heat pumps, RF heating, etc.	• Stable thermal energy • High energy and exergy efficiency • Significant GHG reduction • Applicable in regions with clean grids	• Limited by geographic suitability • High capital costs • Infrastructure readiness required	[3, 9, 39, 40]

facilities in the U.S. alone, producing blades, towers, and generators. As wind turbines grow in size and efficiency, the demand for advanced textile-based composites continues to rise, driving further collaboration between the textile and renewable energy sectors [34, 35]. Overall, wind energy adoption in textiles supports sustainability by providing clean power for manufacturing and enabling the development of eco-friendly materials for the renewable energy industry itself [30, 33, 34].

3.3 Biomass Energy

Biomass energy-based technologies are increasingly explored in textile manufacturing as alternatives to fossil fuels, particularly for process heat and steam generation. Biomass boilers, which utilize agricultural residues (such as rice husks, cotton stalks, and sawdust) or textile biowaste (like cotton apparel waste), are being implemented in several textile-producing regions to address both energy and waste management challenges. Studies on the Portuguese and Sri Lankan textile industries have demonstrated that the use of cotton apparel waste and other biomass sources in boilers is technically and financially feasible, offering reductions in both the cost of steam production and the environmental burden of waste disposal [36, 37]. For example, Sri Lankan apparel factories using cotton waste in biomass boilers have reported decreased reliance on firewood, lower steam production costs, and effective waste valorization, provided that regulatory and technical standards are met [37]. Comprehensive reviews highlight that biomass energy can significantly reduce the carbon footprint of textile manufacturing, especially when agricultural residues are prioritized over dedicated energy crops to avoid land-use change and deforestation [9]. However, sustainability depends on robust supply chain management and monitoring systems to ensure that biomass sourcing does not contribute to ecosystem degradation. While biomass is often positioned as a "green" alternative in the fashion and textile sectors, some reports caution that large-scale adoption of biomass boilers may present risks to climate, ecosystems, and human health if not managed responsibly [38].

3.4 Geothermal and Other Emerging Sources

Geothermal and other evolving technologies are being increasingly accepted as promising solutions for decarbonizing textile production, primarily in regions with access to renewable resources and a need for reliable, low-carbon process heat. Geothermal energy, provides a constant and continuous supply of thermal energy suitable for important textile processes like washing, drying, and dyeing, which consume significant amounts of heat above 100 °C. Recent research has revealed the technical viability and energy efficiency of integrating geothermal systems into textile factories. Cascade utilization systems (CUS) and advanced control algorithms have been recommended to optimize the use of geothermal heat across multiple stages of textile production, improving both energy efficiency and operational flexibility [39, 40]. Energy and exergy analyses of novel integrated geothermal systems for textile factories indicate considerable reductions in both GHG emissions and primary energy consumption in comparison with the conventional fossil fuel-based systems [39]. However, the implementation of geothermal technology is largely dependent on the geographic location of textile plants, as appropriate geothermal resources are not available universally. Substantial capital investment and infrastructure development are also necessary, which can be a barrier in some regions. Beyond geothermal energy, electrification technologies, like electric boilers, industrial heat pumps, and thermal energy storage, are evolving as highly efficient solutions for decarbonizing textile process heat, particularly powered by renewable electricity [9]. These technologies can be incorporated with existing systems and are predominantly promising in countries with decarbonized power grids. Other innovations include radio-frequency heating, vacuum impregnation, and advanced drying techniques, all of which save a considerable amount of energy and also reduce emissions [3]. Therefore, geothermal and other emerging low-carbon technologies present significant opportunities for the textile industry to reduce its carbon footprint, but their deployment depends on local resource availability, infrastructure readiness, and supportive policy frameworks [9, 39, 40]. Figure 1 shows various renewable energy sources used in textile sector.

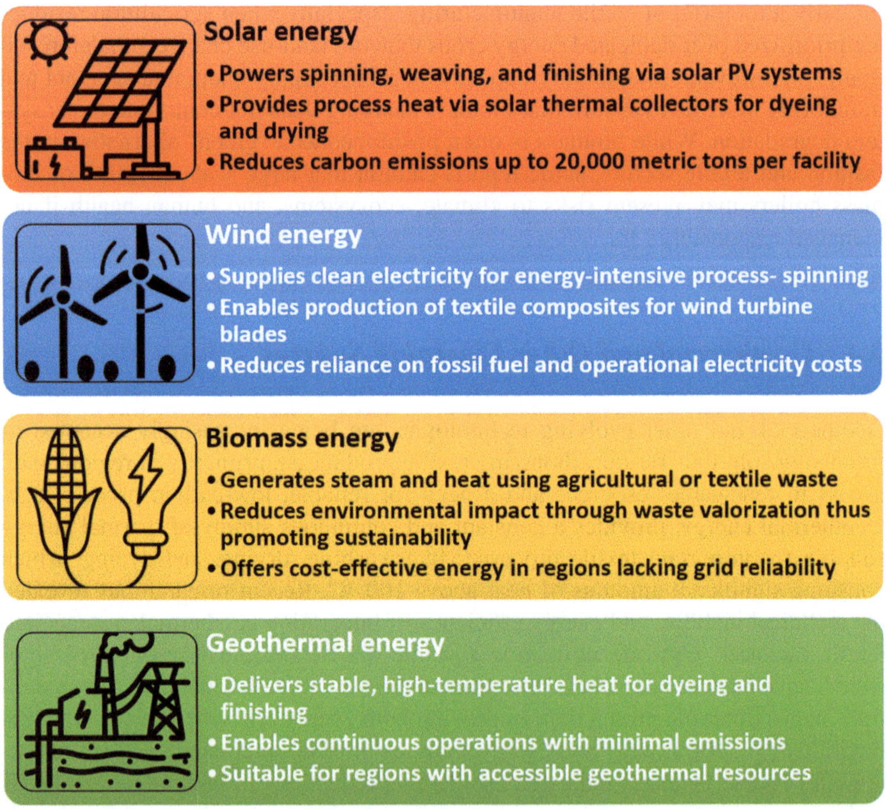

Fig. 1 Renewable energy in textiles. (Source: [36, 39, 41])

4 Technological Innovations for Energy Efficiency

4.1 Energy-Efficient Spinning Machines, Stenter Frames, Dryers

Energy efficiency in textile manufacturing is progressing swiftly through the adoption of innovative spinning machines, dryers and stenter frames, which are amongst the most energy-consuming equipments in the sector. Spinning units solely account for 60–65% of a textile mill's total electricity consumption, making them a main target for efficiency improvements. Research shows that replacing conventional motors with high-efficiency IE4 motors in ring spinning machines can yield annual fund reserves surpassing 24,000 kWh per machine, with repayment periods as short as 1–1.5 years [42]. Further cost-cutting is achieved by optimizing spindle speed curves and using low-friction, lightweight spindles, which reduce peak loads and overall electricity consumption [42, 43]. Stenter frames, crucial for heat-setting and drying fabrics, are also one of the major energy consumers, owing to their high thermal input requirements. Studies show that incorporating heat recovery systems and advanced filtration units can reduce stenter frame energy expenses by 30–35%, while simultaneously alleviating CO_2 emissions and improving process water heating [44, 45]. Case studies from European textile plants reveal that heat recovery from stenter exhaust air can save up to 720 kW per hour, translating to annual cost reductions exceeding €70,000 and CO_2 savings of over 680 tons [44]. Modern industrial dryers equipped with variable frequency drives (VFDs) and optimized airflow systems further contribute to energy savings. Comparative analyses reveal that VFD-equipped dryers consume significantly less electricity than traditional models, even when handling larger air capacities [42]. Collectively, these advancements in spinning, stenter, and drying technologies not only reduce operational costs but also support the textile industry's sustainability and emissions reduction goals [31, 42, 46].

4.2 Smart Grids and Real-Time Monitoring Systems

Smart grids and real-time monitoring systems are revolutionizing energy management and operational efficiency in the textile industry. Unlike conventional grids, smart grids use digital communication, artificial intelligence (AI) and Internet of Things (IoT) sensors to dynamically monitor and optimize electricity consumption pattern across textile manufacturing processes. This enables real-time load balancing, predictive maintenance, and integration with renewable energy sources, considerably improving both energy efficiency and sustainability [47–49]. In practical terms, smart grid pilots in Europe and Asia have demonstrated energy savings of 10–15% in textile dyeing and finishing facilities, where factories can forecast load demands, prioritize low-emission energy sources, and automate alerts for equipment maintenance or anomalies [49]. These systems also enable "demand response" programs, rewarding manufacturers for reducing energy use during peak grid demand, further lowering operational costs and emissions. Real-time monitoring, powered by IoT and Industry 4.0 technologies, enables textile companies to track machine health, predict failures, and optimize production schedules. This results in reduced interruption, lower maintenance costs, and better product quality [48, 50]. Production monitoring systems implemented in small textile companies have led to significant improvements in process control and customer satisfaction [51]. Additionally, smart grid dashboards and APIs facilitate transparent carbon reporting, helping textile manufacturers meet international buyer requirements and regulatory standards for emissions tracking. Initiatives like the Smart Textile Cluster in Portugal exemplify how shared smart grids powered by renewables can allocate energy based on real-time efficiency scores, setting a benchmark for scalable, data-driven sustainability in the sector [49].

4.3 Green Chemistry: Waterless Dyeing, Enzyme Use, Low-Temperature Dyes

Green chemistry innovations are reshaping textile dyeing by promoting waterless processes, enzyme-assisted treatments, and low-temperature dyeing, all aimed at reducing environmental impact and resource consumption. Waterless dyeing technologies, such as supercritical CO_2 dyeing, eliminate the need for large volumes of water and harmful auxiliaries, significantly reducing effluent generation and chemical use. These methods have demonstrated comparable or superior dye uptake and colorfastness to conventional aqueous dyeing, while also lowering energy requirements and process times [52, 53]. Enzyme-based processes are one of the pillars of green chemistry in textiles. Enzymes like cellulases, laccases, and peroxidases, are employed in processes of desizing, scouring, and even dyeing, thus providing high substrate specificity and effective under mild conditions. This reduces the need for harsh chemicals, lowers energy consumption, and minimizes wastewater toxicity. Research highlights that enzymatic treatments can replace conventional alkaline scouring and oxidative bleaching, leading to improved fabric quality and reduced environmental load [54, 55]. Low-temperature dyeing is particularly significant for synthetic fibers like polyester, which traditionally require high temperatures (up to 130 °C) and pressurized equipment. Recent advances in dye chemistry and process engineering have enabled effective dyeing at temperatures as low as 95–100 °C using bio-based auxiliaries and in-situ dye synthesis, achieving similar color strength and fastness properties while reducing energy use by up to 40% [56, 57].

5 Policies and Regulatory Frameworks

5.1 International Directives

The European Green Deal targets to make Europe climate-neutral by 2050, aiming 55% cut in GHG emissions by 2030 in comparison with the 1990 levels [58–60]. The Circular Economy Action Plan (CEAP), a fundamental part of the Green Deal, explicitly addresses textiles as one of the most resource-intensive sectors, promoting sustainable product design, recycling, and resource efficiency [59, 60]. The EU Strategy for Sustainable and Circular Textiles under the CEAP requires that by 2030, textiles on the EU market be durable, recyclable, largely made of recycled fibers, and produced in an environmentally friendly way, with new rules like the Ecodesign for Sustainable Products Regulation (ESPR) and the Digital Product Passport to enhance transparency and traceability [60–62]. Extended Producer Responsibility (EPR) is also being introduced to ensure producers are accountable for the entire lifecycle of textile products, including recycling and reuse [59]. At the global level, the United Nations Framework Convention on Climate Change (UNFCCC) and the Paris Agreement set climate targets that drive national and sectoral decarbonization policies, including for textiles [63, 64]. The UNFCCC's Fashion Industry Charter for Climate Action commits brands to 100% renewable energy in owned operations by 2030 and to phasing out coal-fired boilers in their supply chains [64]. Sustainable Development Goal 7 (Affordable and Clean Energy) calls for increased renewable energy adoption in all sectors, while SDG 13 (Climate Action) urges urgent measures to combat climate change; both are directly relevant to the textile industry's energy transition [65, 66]. These international directives collectively push the textile sector to adopt renewables, improve energy efficiency, and reduce emissions, reinforced by regulatory tools and industry commitments for greater accountability and innovation [60, 62, 64, 65]. A consolidated interpretation of these international policies as well as regulatory frameworks and their implications in textiles is presented in Table 2.

5.2 National Policies Promoting RE in Textiles

National policies in India and China have played a fundamental role in promoting renewable energy adoption within the textile sector. In India, the Perform, Achieve and Trade (PAT) scheme, executed by the Bureau of Energy Efficiency, is a market-based tool targeting energy-intensive industries, like textiles. Under the PAT scheme, designated textile units are given specific energy reduction targets. Units that surpass their targets receive Energy Saving Certificates (ESCerts), which can be traded with those falling short, creating financial incentives for energy efficiency and renewable energy integration [13]. This regulatory structure encourages textile companies for investing in renewable energy sources such as

Table 2 International policies and regulatory frameworks supporting renewable energy and sustainability in textiles

Directive/Agreement	Key objectives	Implications for the textile industry	References
European green Deal	Achieve climate neutrality by 2050; 55% GHG reduction by 2030	Encourages transition to renewable energy and sustainable practices across industries, including textiles	[58–60]
Circular economy action plan (CEAP)	Promote sustainable product design, recycling, and resource efficiency	Targets textiles as high-impact sector; drives circular economy initiatives	[59, 60]
EU strategy for sustainable and circular textiles	Ensure textiles are durable, recyclable, and eco-friendly by 2030	Implements ESPR, digital product passport, and EPR to enhance traceability and lifecycle accountability	[60–62]
UNFCCC and Paris agreement	Set global climate goals and decarbonization pathways	The fashion industry charter urges 100% renewable energy and phasing out coal-fired boilers by 2030	[63, 64]
SDG 7—Affordable and clean energy	Ensure access to sustainable, modern energy for all	Promotes increased use of renewables in energy-intensive sectors like textiles	[65, 66]
SDG 13—Climate action	Urgent actions to combat climate change and its impacts	Drives policy urgency and accountability for energy transition in the textile supply chain	[65, 66]

solar and wind, along with energy-efficient technologies, in order to be compliant with existing policies and save operational costs. In China, the Renewable Energy Law along with a set of environmental regulations have driven the textile industry's transition to cleaner energy. The Renewable Energy Law makes it a mandate for grid companies for purchasing electricity from renewable sources and in exchange provides feed-in tariffs and subsidies for renewable energy projects [67]. Additionally, stringent environmental regulations like Cleaner Production Promotion Law along with heavy environmental taxes, have incentivized textile sector to implement renewable energy and green technologies to curb the emissions and abide by the national standards [68]. Experimental studies approve that these regulatory measures are highly efficient in reducing GHG emissions and PM concentrations, along with supporting technological innovation and enduring sustainability in the sector. These national policies not only drive compliance but also catalyze investment in renewables, placing the textile industry as a leader in industrial decarbonization in both countries [13, 67, 68].

6 Socio-Economic and Community Impacts

6.1 Job Creation in Renewable Energy Installations and Maintenance

Globally, the expansion of renewable energy installations and maintenance has become a major factor in employment generation. The International Labour Organization (ILO) and the International Renewable Energy Agency (IRENA), estimated that around 13.7 million direct and indirect jobs in renewable energy were created worldwide in 2022, with solar PV, wind, and biofuels contributing majorly [69]. Most of these jobs are created during the installation, construction and manufacturing stages, but a significant number are also produced in operations and maintenance (O&M), which offer long-term employment opportunities throughout the lifespan of renewable energy projects [70, 71]. Amongst renewable technologies Solar PV and wind energy, particularly, have the highest job multipliers, which means they generate more employment per unit of energy produced compared to fossil fuels [71, 72]. Solar PV projects are responsible for generating substantial employment while installation and ongoing maintenance phase, with O&M jobs enduring for decades and boosting local economies [71]. Case studies from the World Bank and ILO proved that renewable energy projects not only generate direct jobs in project design, construction and O&M but also encourage indirect employment across entire supply chains and related jobs in allied sectors [69, 70]. In developing countries, the implementation of mini-grid and off-grid solar solutions has led to generation of millions of small and micro enterprises, further increasing employment opportunities [69]. However, the distribution and longevity of these jobs depend on supportive education, skills development, and policy frameworks to ensure a just transition for workers from fossil fuel sectors [69, 70]. Overall, renewable energy installations and maintenance are essential for sustainable job growth during the global energy transition.

6.2 Improved Working Conditions and Community Health Outcomes

The transition to renewable energy in the textile industry is linked to improved working conditions and community health outcomes. Shifting from coal and natural gas to renewables like solar and wind alleviates air pollution levels, lowers the risk of respiratory and cardiovascular diseases among workers and surrounding communities [13]. Research highlights that textile factories powered by renewables emit fewer GHGs and hazardous particulates, directly benefiting both occupational and public health [13, 73]. In India, studies show that integrating decentralized solar power into textile value chains not only reduces environmental impact but also creates safer, cleaner workplaces, especially in rural and semi-urban areas [74].

Improved air quality and reduced exposure to toxic emissions in renewable-powered factories contribute to fewer occupational illnesses and better overall well-being for workers [14, 74]. Additionally, renewable energy adoption often coincides with modernization of equipment and improved factory infrastructure, leading to better ventilation, lighting, and ergonomic conditions [74]. Community health benefits extend beyond factory walls; reduced fossil fuel use means less local air and water contamination, which is particularly important in textile-dense regions where effluent and emissions have historically harmed public health [73, 75]. Furthermore, the creation of green jobs in renewable energy installations and maintenance supports local economies and provides workers with safer, more stable employment opportunities compared to those in polluting industries. Overall, the shift to renewable energy in textiles fosters a healthier, more equitable environment for both workers and communities, aligning with global goals for sustainable development and just transition [14, 74].

7 Barriers to Renewable Energy Adoption in Textiles

7.1 Capital Costs and Financing Limitations

One of the main obstacle to the use renewable energy in textiles is high upfront capital costs especially for small and medium-sized enterprises (SMEs) operating on narrow profit margins. Investments in solar, wind, or biomass systems, as well as the electrification of wet processes, demand significant initial expenditures that many manufacturers cannot afford without external assistance. There still remains a significant gap between the amount of capital required for large-scale decarbonization and current availability despite the emergence of innovative financing models, like blended finance and power purchase agreements (PPAs), at billions of dollars in markets like India [76]. In the absence of substantial subsidies or brand co-investment, renewable projects are not appealing as fossil fuel-based alternatives due to their lengthy payback periods [77]. Furthermore, the challenge for SMEs is exacerbated by the fact that low-interest loans, grants and other financial tools are poorly communicated or limited [12, 78]. Table 3 presents a brief summary of all the barriers of textile industry in adopting renewable energy.

7.2 Technical Integration and Grid Connectivity Issues

Integration of renewable energy with existing infrastructure for textile manufacturing and the broader power grid are the main technical barriers in the systems. Retrofitting is complicated and expensive since many textile facilities have legacy machineries and processes designed particularly for fossil fuels [13, 79]. Grid

Table 3 Barriers to renewable energy adoption in the textile industry

Barriers	Description	Implications for textile industry	References
Capital costs and financing limitations	High upfront investment required for solar, wind, biomass, and electrification technologies. Financing gaps persist, especially for SMEs.	Limits adoption among smaller manufacturers; discourages large-scale decarbonization efforts due to long payback periods and lack of accessible financing.	[12, 76–78]
Technical integration and grid connectivity issues	Existing infrastructure often incompatible with renewables. Grid instability and lack of capacity to manage decentralized systems.	Retrofitting costs are high; unreliable power supply disrupts production; additional costs for storage and hybrid systems increase project complexity.	[9, 13, 79, 80]
Knowledge and awareness gaps	Limited awareness of technical, financial, and environmental benefits of renewable energy. Lack of access to tools and training.	Leads to hesitation and resistance in adopting clean technologies; poor understanding of ROI and compliance benefits.	[78, 81, 82]
Policy fragmentation and industry inertia	Inconsistent regulations, fossil fuel subsidies, and absence of renewable energy mandates. Resistance to innovation within the industry.	Reduces competitiveness of renewables; fragmented efforts delay progress; lack of coordinated action stifles investment and policy alignment.	[9, 13, 79]

infrastructure in major textile-producing countries may be unreliable, underdeveloped, or insufficient to accommodate dispersed renewable energy inputs on a large scale, causing concerns about power quality and continuity [9]. Vigorous grid connectivity and, in some situations, significant improvements to local distribution networks are necessary for the electrification of thermal processes. Furthermore, there is a rise in complexity and cost, because of the intermittent nature of solar and wind energy which necessitates investments in energy storage or hybrid systems [9, 80].

7.3 *Knowledge and Awareness Gaps*

The lack of knowledge and awareness amongst textile manufacturers about the advantages, technical specifications and funding sources of renewable energy constitutes a major barrier [78, 82]. Many decision-makers and factory managers are unknown and unaware about the energy efficiency benchmarks, available technologies, or the commercial case for renewables. This is restricted by market information, limited access to qualified assessors and resource assessment tools, causing risks of aversion and inertia [78, 81]. Information sharing and best practices are further hindered by communication gaps amongst technology providers, financial institutions, and manufacturers. Regular training sessions, clear case studies and demonstration projects, are insufficient, resulting in a perception that renewable energy projects are high-risk or not aligned with core business priorities [78].

7.4 Policy Fragmentation and Industry Inertia

Another systemic barrier to renewable energy adoption is the policy fragmentation and industry inertia. Unreliable policy frameworks, fossil fuel subsidies and inefficient policy execution in many countries, hamper the competitive ability of renewables [13]. The textile industry, characterized by a disseminated and highly competitive supply chain, often lacks advocacy for policy change or coordinated action. Rather than technological innovation, many leading brands and manufacturers have conventionally placed a higher priority on cost reduction via relocation. This leads to a culture that is resistant to disruptive change. It is challenging for companies to justify investments in clean energy for longer term because of the absence of transparent, legally binding regulations for utilization of renewable energy and the lack of regionally uniform standards [13, 79]. The inefficient government action on grid modernization, emissions regulations and renewable energy incentives further exacerbate the problem, forcing individual manufacturers to implement a contradictory and complex policy landscape [9].

8 Strategies for Scaling Renewable Integration

8.1 Holistic Planning and Energy Audits

An initial step for effective renewable energy incorporation is all-inclusive energy planning, reinforced by widespread energy audits. Current energy utilization patterns and identification of inefficiencies are thoroughly assessed in energy audits along with highlighting opportunities for renewable incorporation modified for specific processes like spinning, dyeing, weaving, and finishing [6]. Studies in Pakistan and Bangladesh highlight that optimal utilization of sustainable energy in textiles is only made practical by proper planning and comprehensive energy audits, which help determine the efficiency measures for each facility and the best combination of renewable sources [6, 83]. The entire energy system of a textile plant, including process heat, power, and the potential of waste heat recovery, is taken into account during holistic planning. With this approach, operational expenses are maintained which enables savings and emissions are reduced through the integration of renewable energy sources. Practically, this would imply incorporating solar PV for electricity generation, solar thermal to heat processes, and storage-enabled hybrid systems for addressing intermittency [6]. Furthermore, energy audits provide data required for support the funding or incentive applications, validate investments and recommend technology selection. Table 4 gives a detailed information of various strategies for expansion of renewable energy.

8.2 *Public-Private Partnerships*

Public-private partnerships (PPPs) are gaining impetus and recognition as powerful tools for accelerating sustainable approaches in textiles, which includes renewable energy adoption. PPP includes collaboration of government organisations and private companies' stakeholders for financing, implementation and management of renewable energy projects. Governments aid support through its policy frameworks, infrastructure, and financial assistance, while private sector contributes operational proficiency, investment capital and technological innovation [84]. By combining resources and dividing risks, PPPs have an ability to overcome barriers like significant upfront expenses and fragmented regulatory environments. They are especially effective in extensive projects like grid modernization initiatives, community solar

Table 4 Strategies for expansion of renewable energy

Strategies	Purpose and description	Key benefits for the textile sector	References
Holistic planning and energy audits	• Conduct detailed assessments of energy usage and inefficiencies to design integrated renewable energy systems (e.g., PV, solar thermal, hybrid systems).	• Identifies tailored renewable solutions • Justifies investment with data • Enables optimal mix of technologies • Enhances operational reliability	[6, 83]
Public-private partnerships (PPPs)	• Establish collaboration between governments and private firms to finance and implement renewable projects	• Reduces financial and operational risk • Enables large-scale infrastructure • Supports policy alignment and innovation	[83, 84]
Renewable energy purchase agreements (PPAs)	• Long-term contracts for procuring renewable electricity from power producers (onsite/offsite).	• Minimizes upfront capital investment • Offers price stability • Helps meet regulatory and climate targets	[13, 85]
Training and capacity building	• Equip workers and managers with skills to operate, maintain, and manage renewable systems and sustainability practices.	• Ensures safe and efficient operations • Supports informed decision-making • Accelerates technology adoption • Enhances job creation and transition readiness	[12, 14, 83]

parks and shared biomass facilities, which are otherwise challenging for individual firms to manage alone [84]. Collaborative efforts in Bangladesh, amongst government, private companies and international organizations have proven to be essential for creating a supportive environment for renewable energy in textile sector [83]. PPPs also play a role in scaling up textile recycling and circular economy initiatives, which are increasingly linked to renewable energy adoption [84].

8.3 Renewable Energy Purchase Agreements (PPAs)

Renewable energy purchase agreements (PPAs) are contracts that facilitate textile manufacturers for acquiring renewable electricity directly from autonomous electricity providers, often at fixed rates for prolong periods [13, 85]. PPAs enable companies to acquire clean energy without any need to purchase generating assets with a substantial upfront capital investment. They can be either structured as onsite like rooftop solar or offsite like utility-scale solar/wind farms. PPAs are particularly useful in markets with limited renewable energy supplied by the grid or where direct acquisition is encouraged by regulatory frameworks. They assist businesses in meeting regulatory requirements as well as sustainability goals, stabilize prices and protect against future energy cost fluctuations [13, 85]. PPAs have been adopted by well-known global brands like Nike and H&M to advance their renewable energy goals, and emerging markets like Pakistan and Vietnam are adopting comparable models [13]. Supporting policies, clear legal frameworks and the availability of reliable renewable energy sources are necessary for PPA success.

8.4 Training and Capacity Building for Workers and Managers

The successful incorporation of renewable energy technologies in textile industry depends on the skills as well as knowledge of the workforce and management. Training and capacity building are vital for ensuring that employees can maintain as well as operate new systems safely and effectively, and that managers comprehend the renewable energy business case along with financing and regulatory procedures [12, 14]. Regular training programs addressing technical aspects (like energy management systems, operation of solar PV and biomass boilers), safety protocols, and energy efficiency best practices are necessary [14]. Capacity building should also focus on decision-makers, by equipping them with the resources for conducting cost-benefit analyses, assessment of project risks, and involvement with government or financial institutions [12]. Peer learning networks and demonstration projects can accelerate technology implementation and promote an innovative culture in developing countries, where there are knowledge and awareness gaps [83]. As the industry shifts away from fossil fuels International organizations like ILO emphasize that training in renewable energy not only enhances operational outcomes but also generates employment [14]. Capacity building should be incorporated in broader sustainability strategies, in order to benefit from renewable energy including emissions reductions, cost savings and better working conditions.

9 Conclusion and Future Prospects

The textile industry's shift towards renewable energy offers numerous advantages and opportunities. Textile manufacturers can substantially reduce their GHG emissions, mitigate climate hazards, and efficiently safeguard energy by replacing fossil fuels with renewable energy forms like solar, wind, biomass and geothermal energy. Along with reducing environmental impact, renewable technologies also enhance process efficiency and operational cost-efficiency. Solar thermal systems can cater up to 50% of a (textile manufacturing) plant's heating requirements, whereas solar PV installations on the rooftop can considerably reduce electricity expenses. Similarly, wind power links the textile sector with broader renewable energy value chains by serving as a clean energy source and by innovation of sustainable materials for components used wind turbines. The urgent need for climate action makes decarbonization necessary across all industrial sectors and textiles which account for almost 10% of global GHG emissions. The global climate goals cannot be achieved with continued dependency on natural gas and coal, particularly in major textile hubs like India, China, and Bangladesh. Because of emerging stringent global sustainability regulations, prompt action for decarbonization through the integration of renewable energy is necessary for both environmental concern as well as for preserving industry competitiveness. By strategically implementing renewables in combination with energy efficiency initiatives, green chemistry advances and smart grid technologies, textile industries can successfully achieve zero-carbon goals. The implementation of electrified systems driven by renewable energy, along with low-carbon substitutes for thermal processes, provides a clear pathway. Although empowering this vision will require significant investment in infrastructure, innovation and skills development, but the long-term advantages will exceed the initial costs. Both, international and local co-ordinated systems are essential for accomplishing this transformation. International policy frameworks including, Paris Agreement and Sustainable Development Goals (SDGs) provide a guiding structure, whereas national policies and laws like Renewable Energy Law of China and Perform, Achieve and Trade (PAT) scheme from India are essential enablers. However, co-operation and co-ordination between governments, private businesses, financial institutions, and civil society is required for actual development. Energy purchase agreements (PPAs), Public-private partnerships (PPPs) and industry-led sustainability commitments require expansion. Simultaneously, education, awareness and capacity-building programs are essential for equipping managers and employees with the knowledge as well as skills for successful management of energy transition. In conclusion, incorporating renewable energy within textiles has the potential for redefining the industry's future along with being environmental imperative. The transition safeguards climate resilience in addition to economic benefits, and better social outcomes.

References

1. Bravo MVC, Iturralde WMP (2022) Energy efficiency in the competitiveness of the textile industry. Open J Bus Manag 10(4):1755–1767
2. Gunturu KPK, Kota KK, Sharma M (2022) Energy efficiency improvement opportunities in Indian textile industries. Text Leather Rev 5:296–326
3. Amrutrao A, Degnet DD, Gota GY (2020) Use of energy in textile industry: a case study on Ethiopian textile industry. IOSR-JMCE 17(3, Ser. I):58–64
4. Imran S, Mujtaba MA, Zafar MM, Hussain A, Mehmood A, Farwa UE, Saleel CA (2023) Assessing the potential of GHG emissions for the textile sector: a baseline study. Heliyon 9(11):e22404
5. Leal Filho W, Perry P, Heim H, Dinis MAP, Moda H, Ebhuoma E, Paço A (2022) An overview of the contribution of the textiles sector to climate change. Front Environ Sci 10:973102
6. Kousar S, Shafqat U, Kausar N, Pamucar D, Karaca Y, Salman MA (2022) Sustainable energy consumption model for textile industry using fully intuitionistic fuzzy optimization approach. Comput Intell Neurosci 2022(1):5724825
7. GCI (Global Climate Initiatives) (2024). The carbon impact of the textile industry: assessment and sustainable solutions. https://globalclimateinitiatives.com/en/limpact-de-lindustrie-du-textile-bilan-et-solution-durable/
8. Leal Filho W, Dinis MAP, Liakh O, Paço A, Dennis K, Shollo F, Sidsaph H (2024) Reducing the carbon footprint of the textile sector: an overview of impacts and solutions. Text Res J 94(15–16):1798–1814
9. Hasanbeigi A, Springer C, Wei D (2024) Low-carbon thermal energy technologies for the textile industry. Global Efficiency Intelligence. https://www.globalefficiencyintel.com/lowcarbon-thermal-energy-technologies-for-the-textile-industry
10. Silva B, Malheiro D, Júnior D, Nunes F, Santos J, Guimarães L, Sousa RM (2025) Analysis of the carbon footprint of a textile company for the automotive industry. Energies 18(10):2478
11. Sharma BM (2024) Transforming textile manufacturing: the impact of renewable energy and efficiency. TheCSRUniverse. https://thecsruniverse.com/articles/transforming-textiles-rswm-s-path-to-a-greener-future
12. Pawar N (2024) Embracing renewable energy: a guide for Indian textile MSMEs. Neufin. https://neufin.co/blog/renewable-energy-advantage-for-textiles-india/
13. Buckley S (2022) The textile industry's transition to renewable energy. Fibre2Fashion. https://www.fibre2fashion.com/industry-article/9496/the-textile-industry-s-transition-to-renewable-energy
14. ILO (International Labour Organization) (2022) A just transition in the energy, textile & garment sectors in Indonesia: Technical stakeholder workshop, 24–25 October 2022, Jakarta. https://www.ilo.org/resource/just-transition-energy-textile-garment-sectors-indonesia-technical
15. Kabiraj M (2022) How clean energy could shape India's textile industry: challenges and opportunities. Council on Energy, Environment and Water (CEEW). https://www.ceew.in/blogs/how-clean-renewable-energy-could-solve-challenges-in-indian-textile-industry
16. Hossain MA, Samanta AK (2019) A cost minimization process of heat and energy consumption for direct dyeing of cotton fabric coloration with triethanolamine. J Text Eng Fash Technol 5(5):235–240

17. Bureau of Energy Efficiency (2010). Detailed project report on installation of condensate recovery pump in jet dyeing machine (Pali textile cluster). https://www.sameeeksha.org/pdf/dpr/PAL_TXT3.pdf
18. Yalçın T, Arab G, Şener A, Gördebil K, Akçal E (2020) Energy efficiency in textile industry for small and medium-sized businesses. Appl Res Stud Pract 16(1):21–37
19. Quantis (2018) Measuring fashion: environmental impact of the global apparel and footwear industries study—Full report and methodological considerations. https://quantis-intl.com/report/measuring-fashion-report/
20. Guo Y, Zhu L, Wang X, Qiu X, Qian W, Wang L (2022) Assessing environmental impact of NOX and SO2 emissions in textiles production with chemical footprint. Sci Total Environ 831:154961
21. Dhameliya KB, Ambasana C (2023) Assessment of wastewater contaminants caused by textile industries. J Pure Appl Microbiol 17(3):1477–1485
22. Butt MS (2021) Replacing fossil fuel-based thermal generation with renewable fuel in the textile and garments sector in Bangladesh: prefeasibility study (final V-03). Espire Consult Pakistan. https://www.espire.com.pk
23. Dominish E, Sharpe S, Martínez-Fernández C (2022) Taking climate action: measuring carbon emissions in the garment sector in Asia. International Labour Organization
24. Khuky MA, Hook LS, Chin L, Saari MYB (2022) The impact of textile and clothing export on environmental quality in Bangladesh: an ARDL bound test approach. Nat Environ Pollut Technol 21(3):1421–1430
25. Ozturk HK (2005) Energy usage and cost in textile industry: a case study for Turkey. Energy 30(13):2424–2446
26. Hasanbeigi A, Springer C (2024) Electrification of heating in a textile plant in India: a case study. Global Efficiency Intelligence. https://www.globalefficiencyintel.com/electrification-of-heating-in-a-textile-plant-in-indiacase-study
27. Naqvi SM, Bakhtiar A, Mahachandra M (2024) Energy audit of a textile company: a case study of finishing textile. World J Adv Res Rev 22(3):1040–1047
28. Apparel Resources (2020) Powering the apparel and textile industry with solar energy: Tamil Nadu the frontrunner. Apparel Resources. https://apparelresources.com/business-news/sourcing/powering-apparel-textile-industry-solar-energy-tamil-nadu-frontrunner/
29. Sadman SAM (2025) Solar power solutions for the textile industry: weaving a greener future. Textile Research and Development. https://textilernd.com/solar-power-solutions-for-the-textile-industry-weaving-a-greener-future/
30. JMK Research (2024). Green power procurement in the Indian textile sector: case study of Arvind Ltd. https://jmkresearch.com/renewable-sector-published-reports/green-power-procurement-in-the-indian-textile-sector/
31. Tanim MAM (2024) Solar thermal solutions for textile industry's net-zero goal. TexSpace Today. https://www.texspacetoday.com/solar-thermal-solutions-for-textile-industrys-net-zero-goal/
32. Sharma M, Tyagi VV, Chopra K, Kothari R, Singh HM, Pandey AK (2023) Advancement in solar energy-based technologies for sustainable treatment of textile wastewater: reuse, recovery and current perspectives. J Water Process Eng 56:104241
33. O'Mahony M (2021) Textiles' role in a growing energy industry. Textile Technology Source. https://textiletechsource.com/2021/06/14/textiles-role-in-a-growing-energy-industry/
34. Lucintel (2024). Woven textile in wind energy market report: trends, forecast and competitive analysis to 2030. https://www.giiresearch.com/report/luci1616957-woven-textile-wind-energy-market-report-trends.html
35. DOE (U.S. Department of Energy) (2025) Wind manufacturing and supply chain. U.S. Department of Energy. https://www.energy.gov/eere/wind/wind-manufacturing-and-supply-chain
36. Nunes LJR, de Oliveira Matias RGJC (2019) Technological innovation in biomass energy for the sustainable growth of textile industry. Sustainability 11(2):528
37. Prasanga WC, Ranathungage AN, Rathnasiri P (2021) Use of cotton apparel waste as an energy source for biomass boilers: a feasibility study. Int J Des Nat Ecodyn 16(1):41–51

38. Zhang X, Joshi S, Cook G, Kitchin R, Ren S (2023) Biomass burning: the fashion industry's false phase-out. Stand.earth. https://stand.earth/resources/biomass-burning-the-fashion-industrys-false-phase-out/
39. Khanmohammadi S, Saadat-Targhi M, Nabati A (2022) Energy and exergy analyses of a new integrated system for textile factory using geothermal energy source. Energy 257:124696
40. Saadat-Targhi M, Khanmohammadi S (2025) A sustainable multi-level textile thermal energy system integrated with geothermal energy source: thermal simulation and multi-objective optimization. J Environ Chem Eng 13(3):116647
41. Satharasinghe A, Hughes-Riley T, Dias T (2020) A review of solar energy harvesting electronic textiles. Sensors 20(20):5938
42. Qaisar MH, Khattak A, Bilal M, Imran K, Ulasyar A (2021) Electrical energy Management in Spinning Area of textile industry. In: In 2021 4th international conference on energy conservation and efficiency (ICECE). IEEE, pp 1–6
43. Goyal A, Kumar D (2013) Energy-efficiency improvement in the spinning industry. Int J Eng Res Technol 2(6):1–7. https://www.academia.edu/25448337/Energy_efficiency_improvement_in_spinning_industry
44. KMA Umwelttechnik GmbH (2021) Heat recovery at the stenter frame to reduce operating costs and CO_2 emissions. Text Mag. https://www.kma-filter.com/industries/case-studies-overview/example-stenter-frame/
45. Vanterm (2019). Energy saving opportunities for stenter frames. https://vanterm.com/energy-saving-opportunities-for-stenter-frames/
46. Ozturk E, Cinperi NC, Kitis M (2020) Improving energy efficiency using the most appropriate techniques in an integrated woolen textile facility. J Clean Prod 254:120145
47. Kathirvel M, Chandrasekaran M (2025) Predictive maintenance and energy optimization with AI-driven IoT framework in textile manufacturing industry. Int J Comput Exp Sci Eng 11(2):2408–2422
48. Shah V, Singh S, Taur S (2025) Impact of IoT and industry 4.0 on textile industry and how it optimizes the process. Int Res J Mod Eng Technol 7(3):1–10
49. TexSpace Today (2025) Smart grids bring precision power to manufacturing. https://www.texspacetoday.com/smart-grids-bring-precision-power-to-manufacturing/
50. Petrillo A, Rehman M, Baffo I (2024) Digital and sustainable transition in textile industry through internet of things technologies: a Pakistani case study. Appl Sci 14(13):5380
51. Trisdiyanti R, Renaldi F, Umbara FR, Bon AT (2020) Implementation of production monitoring systems in a small textile company. In: Proceedings of the 5th North American international conference on industrial engineering and operations management. IEOM Society International, pp 10–14
52. Pandit P, Singha K, Maity S, Ahmed S (eds) (2022) Textile dyes and pigments: a green chemistry approach. Wiley
53. Roy Choudhury AK (2013) Green chemistry and the textile industry. Text Prog 45(1):3–143
54. Raj A, Chowdhury A, Ali SW (2022) Green chemistry: its opportunities and challenges in colouration and chemical finishing of textiles. Sustain Chem Pharm 27:100689
55. Sehrawat A (2023) Applications of green chemistry principles in textile wet processing. J Adv Sci Res 14(09):1–5
56. Li S (2018) Low temperature dyeing of polyester Fibres through in-situ dye synthesis. (Doctoral dissertation). University of Leeds
57. Radei S (2020) Improvement of polyester dyeing at low temperature. (Doctoral thesis). Universitat Politècnica de Catalunya
58. DuPont (2024) The European Green Deal: how it is reshaping the future of flame-resistant (FR) textiles [White paper]. https://www.dupont.com/personal-protection/nomex-comfort-with-ecoforce-technology.html
59. Fibre2Fashion (2024). Green policies in the EU: advancements and impacts. https://www.fibre2fashion.com/industry-article/10162/green-policies-in-the-eu-advancements-and-impacts

60. Lüttin L (2025) Full overview of the EU textile strategy and regulations. Carbonfact. https://www.carbonfact.com/blog/policy/eu-regulations-for-textile-brands
61. European Commission (2022) EU strategy for sustainable and circular textiles. https://environment.ec.europa.eu/strategy/textiles-strategy_en
62. European Commission (2025) EU strategy for sustainable and circular textiles [Legislative proposal]. https://www.europarl.europa.eu/legislative-train/theme-a-european-green-deal/file-eu-textiles-strategy
63. Textile Exchange. (2022) Climate Impact Module Companion Guide (Version 1.0). https://textileexchange.org/app/uploads/2022/08/Climate-Module-Companion-Guide-.pdf
64. UNFCCC (United Nations Framework Convention on Climate Change) (2021) Fashion industry charter for climate action. UNFCCC. https://unfccc.int/climate-action/mou/fashion-industry-charter
65. KPMG & Textile Exchange (2022) Threading the needle: weaving the sustainable development goals into the textile, retail, and apparel industry. Textile Exchange. https://textileexchange.org/knowledge-center/documents/threading-the-needle-report/
66. Thakker AM, Sun D (2023) Sustainable development goals for textiles and fashion. Environ Sci Pollut Res 30(46):101989–102009
67. Guérin, E., & Schiavo, J. (2011). Chinese renewable energy and technology policies: legal compatibility with WTO rules & economic interactions with other countries' climate and industrial policies. IDDRI Paper (02/11). https://www.iddri.org/en/publications-and-events/iddri-papers/chinese-renewable-energy-and-technology-policies-legal-compatibility
68. Bibi S, Khan A, Fubing X, Jianfeng H, Hussain S, Khan AN (2024) Impact of environmental policies, regulations, technologies, and renewable energy on environmental sustainability in China's textile and fashion industry. Front Environ Sci 12:1496454
69. IRENA (International Renewable Energy Agency), ILO (International Labour Organization) (2023) Renewable energy and jobs: annual review 2023. International Renewable Energy agency/International Labour Organization, Abu Dhabi/Geneva
70. Dobrotková Z, Pargal S, Fichaux N (2024) Jobs for a livable planet: job creation potential of the clean energy transition. Energy Sector Management Assistance Program (ESMAP), World Bank. https://documents.worldbank.org/en/publication/documents-reports/documentdetail/099012324075618467/p17643810e6ae80901a66b1c56cb14b2a36
71. Kim J, Mohommad MA (2022) Jobs impact of green energy. International Monetary Fund. https://www.imf.org/en/Topics/Climate-Change/jobs-impact-of-green-energy
72. Ram M, Aghahosseini A, Breyer C (2020) Job creation during the global energy transition towards 100% renewable power system by 2050. Technol Forecast Soc Chang 151:119682
73. Rahaman MT, Pranta AD, Repon MR, Ahmed MS, Islam T (2024) Green production and consumption of textiles and apparel: importance, fabrication, challenges and future prospects. J Open Innov Technol Mark Complex 10(2):100280
74. UN PAGE (United Nations Partnership for Action on Green Economy) (2023) Assessment of green jobs and decent work opportunities in the textile/garment sector in Jharkhand State, India. https://www.un-page.org/2023-annual-report/assessments/
75. European Parliament. (2020). The impact of textile production and waste on the environment (infographics). https://www.europarl.europa.eu/topics/en/article/20201208STO93327/the-impact-of-textile-production-and-waste-on-the-environment-infographics
76. Jawale L, Perkins L (2025) Decarbonizing India's textile industry: a US$6 billion climate finance opportunity. Impact Entrepreneur. https://impactentrepreneur.com/decarbonizing-indias-textile-industry/
77. Indian Textile Magazine (2023) Indian textile companies harnessing solar power for sustainable growth. https://www.indiantextilemagazine.in/indian-textile-companies-harnessing-solar-power-for-sustainable-growth/
78. ISC (Institute for Sustainable Communities), ERI (Engineering Resources International) (2020) Clean technology adoption for textile SMEs in Bangladesh—Final report. United

States Agency for International Development (USAID). https://sustain.org/wp-content/uploads/2020/10/CleanTech-Final-Report-ERI-ISC.pdf
79. Hasan AM, Rokonuzzaman M, Tuhin RA, Salimullah SM, Ullah M, Sakib TH, Thollander P (2019) Drivers and barriers to industrial energy efficiency in textile industries of Bangladesh. Energies 12(9):1775
80. Thermax Limited (2024) Navigating energy transition: comprehensive solutions for textile players in the pursuit of sustainability. https://www.thermaxglobal.com/navigating-energy-transition-comprehensive-solutions-for-textile-players-in-the-pursuit-of-sustainability/
81. Alam MFB, Hosen MI, Mridha JH, Chowdhury SE, Rahman MA (2023) Assessing the barriers of integrating technological innovations in textiles sector: implications towards sustainable production. Green Technol Sustain 1(3):100039
82. Khan MMA, Alam MJ, Saha S, Sayem A (2024) Critical barriers to adopt sustainable manufacturing practices in medium-sized ready-made garment manufacturing enterprises and their mitigation strategies. Heliyon 10(20):e39195
83. Noor M, Soad SP, Ahmmed MM (2024) Renewable energy integration in Bangladesh's textile industry: analyzing opportunities amidst challenges. American International University—Bangladesh, Kuratoli. https://www.aiub.edu/Files/student-research/Renewable_Energy_Integration_in_Bangladeshs_Textile_Industry_Analyzing_Opportunities_Amidst_Challenges.html
84. Cui C, Liu Y, Xia B, Jiang X, Skitmore M (2020) Overview of public-private partnerships in the waste-to-energy incineration industry in China: status, opportunities, and challenges. Energ Strat Rev 32:100584
85. Ecohz. (2021). How clothes and shoes can be made with renewable electricity. https://www.ecohz.com/sustainability-solutions/clothes-and-shoes

Renewable Energy in the Indian Textile Sector

Madhurima S. Rudra and Subramanian Senthilkannan Muthu

Abstract The textile industry is one of the most energy intensive sectors globally. A large part of the energy needs in this sector is conventionally sourced through electricity and thermal pipelines. With increasing climate awareness and the need to reduce carbon emissions, the sector is undergoing transformation by adopting renewable energy sources such as solar, wind and hydropower. This chapter explains the energy requirement in the textile manufacturing process for various types of yarn and fibres, and the need to shift to renewable options not just as a measure of green initiative but also to boost the quantity and quality of production significantly. The chapter highlights the textile industry in India—one of the largest textile manufacturers in the world and explores the cost benefit analysis between traditional and renewable energy sources.

Keywords Textile industry · Renewable energy · Greenhouse gas emissions · Sustainability · Fossil fuels · Energy transition · Environmental impact

1 Introduction

The textile industry was the first sector to be industrialised during the revolution and since then, has undergone multiple developments nearly replacing entire manpower with machines across different stages of production. The industry is also highly chemical intensive, using harmful substances such as dyes that lead to toxic waste generation. The dual problem of increased fossil fuel consumption and increased waste generation can be mitigated using renewable sources of energy.

M. S. Rudra
MOP Vaishnav College for Women, Chennai, Tamil Nadu, India

MOP Vaishnav College for Women, Chennai, Tamil Nadu, India

S. S. Muthu (✉)
Green Story Inc., Tung Chung, Hong Kong

S. S. Muthu (ed.), *Renewable Energy Implications for the Textile Industry*, Sustainable Textiles: Production, Processing, Manufacturing & Chemistry,
https://doi.org/10.1007/978-3-032-10917-0_6

The Indian Textile Industry is one of the most prominent manufacturing hub in the subcontinent. As the global leader in cotton, jute, and silk production, India caters to an international market of US $37.7 Bn and a domestic market of US $142.5 Bn [1]. The primary sources of energy for the sector's manufacturing are through central electricity grid systems and thermal energy using coal and petroleum products. Due to increasing climate campaigns and stricter environmental regulations, the textile industry is pressured to opt for sustainable operations. While governments impose carbon bans, consumers also prefer eco-friendly products, making the transition to cleaner fuels of great strategic importance.

On the profit side, energy costs contribute a significant amount to the production expenses. Rising uncertainty in oil and natural gas supply and concerns regarding energy security are prompting manufacturers to look for renewable alternatives. Advancements in technology have enabled sustainable solutions such as rooftop solar systems, wind turbines, and biomass boilers that are becoming more accessible and affordable for textile manufacturers of all sizes.

The chapter highlights the Indian Textile industry, its distribution, and trends that set the context for understanding the significance of energy consumption in the sector. It further explores the different sources of renewable energy, such as solar, wind, green hydrogen, and hydropower as replacements for the conventional electricity and thermal energy systems. The chapter also provides an in depth analysis of the Decentralised Renewable Energy initiative by the Government of India for promoting customised energy sources for textile hubs rather than dependence on central gridlines. Further sections explore the cost benefit analysis of adopting renewable energy sources against conventional energy and the challenges that lie ahead for renewable energy systems in the Indian textile market.

2 Textile Industry in India

2.1 Market Overview

The Textile Industry is one of the oldest sectors in India, dating back to the Indus Valley civilisation. The rich culture and vibrant heritage of the country have given rise to traditional fabrics and clothing that are unique to different parts of India. From pashmina in Kashmir to silk in Kancheepuram, the varied yarns, along with the differences in draping styles, create several opportunities for garment manufacturers to cater to. The Textile Industry in India contributes 4.1% of the global textile and apparel trade with a projected market size of $350 Bn by 2030 [2]. The sector employs 45 million people, from traditional hand-spun and hand-woven textiles to mass-manufacture mills. The global apparel market is expected to grow at a CAGR of 8.9% and is estimated to reach US $65 Bn by 2026 [2].

The production of textiles in India is structured into organised, unorganised, and decentralised sectors. The organised industry primarily consists of spinning and composite mills, while the unorganised sector caters to the handloom, powerloom, and fabric processing units. The decentralised handloom, powerloom, and hosiery units form the largest component of the textile industry. This sector is characterised by small and medium enterprises primarily dealing in cotton yarn.

2.2 *Trends in the Textile Industry*

The textile and apparel industry in India has strengths across the value chain from fibre and yarn to fabric and apparel. The rising population and demand for consumer choices remain the primary drivers for revenue growth in the sector. Amid concerns regarding fast fashion and sustainability, many producers have opted for organic fabrics and renewable manufacturing processes, incorporating waste recycling and circular methods of production. There is a high demand for natural and manmade fibres stimulated globally. The natural fibres market is expected to grow from US $138 Bn to US $195 Bn by 2026 [2], with India being a solid competitor in the raw material sector for cotton, silk and woollen yarns. The sector benefits from the government through its various schemes and policies, such as National Handloom Development Programme (NHDP) and PM Mega Integrated Textile Region and Apparel (PM—MITRA) Parks across the country. The Technical Textile sector is also a rapidly increasing market for industrial, medical and sport apparel needs. The technical textile sector combines the creativity of fabrics with technology to produce materials with exceptional insulation and heat resistance that are valuable to the automotive, healthcare and eco-friendly markets.

2.3 *Regional Distribution*

India is a vast country with diverse textile crafts and fabrics. This has led to the customisation of clothing, acclimatising to the region's geography. The majority of the country's cotton production is concentrated in the states of Maharashtra and Gujarat, which also form the base for the subsequent processing methods. Tamil Nadu is also known for its massive cotton production and textile hubs such as Coimbatore and Tirupur in the state. West Bengal has emerged as the largest jute producing region in the world due its suitable climate and marshy soil. Subsequently, a majority of the jute production is concentrated in the eastern and north-eastern states of India.

Metropolitan cities like Mumbai, Ahmedabad and Chennai are important centres for textile and apparel manufacturing and shipping in the country. Apart from these cities, every region has its own fabric, weaving design or knitwear, such as pochampally, ikkat and kalamkari inspired by folklore and passed down traditions through the generations.

2.4 Major Natural Textile Types

The textiles in India can be categorised mainly into natural and manmade textiles. There is a large domestic as well as export market for both materials, making India the fifth largest exporter of textiles and apparel in the world. Natural fibers such as cotton, silk, jute, and wool generated a revenue of US $5047.2 million in 2024 for India and are projected to grow at a CAGR of 6.3% till 2030 [3].

2.4.1 Cotton

As the second largest producer of cotton in the world, India leads the cotton yarn segment, contributing to 23% of the global cotton production [2]. The country's cotton production is divided into Central Zone (Maharashtra, Gujarat, Madhya Pradesh), Northern Zone (Punjab, Haryana and Rajasthan) and Southern Zone (Telangana, Andhra Pradesh and Karnataka) that jointly produce an estimated 302.25 lakh bales or 170 kg [4]. Electric energy is the primary source of energy in the cotton industry. A study reports 93% of energy used in cotton spinning mills, 85% in the weaving mills and 65% in clothing companies is electrical energy [5]. The Cotton industry operates both in the organised and unorganised spheres. The majority of spinning and weaving mills form part of the organised cotton industry, while handloom, powerloom form the base of small and medium-sized enterprises in the unorganised sector.

2.4.2 Silk

Sericulture in India has a rich history and many regional variations, producing a multitude of silk fabrics. India is the second-largest producer of silk fabric and garments in the world. It produces all four varieties of natural silk—Mulberry, Eri, Tasar, and Muga. The silk industry in India is overseen by the Central Silk Board and is energy intensive in all the stages of production—cocoon drying, cooking, reeling and re-reeling. For every 60 kg of raw silk produced, about 1000 kg of firewood is required in the conventional production process [6]. Modifications in the production have led to the implementation of solar and biomass energy sources to reduce the dependence on firewood as the sole energy generator in the sector.

2.4.3 Jute

The Jute industry plays a significant role in the Indian economy, employing over 3.5 million workers, producing 358 lakh tonnes of hessian, sacking, food-grade, carpets, and decorative jute fabrics [7]. While India produces 70% of the world's jute production, 90% of it is used for domestic consumption due to high demand in the country. There has been a steady growth of jute exports over the years especially since jute stands as a eco friendly and safe packaging alternative against plastics. The jute manufacturing process is extensively dependent on electric power from softening the raw jute to carding, drawing and spinning to get the jute yarn for sale as well as an input to make other jute products.

2.4.4 Wool

India produces three categories of wool—Carpet Grade, apparel grade and coarser grade that are split into organised and the decentralised sectors. The majority of the wool produced is for the export trade, comprising handmade carpets, winter wear, knitwear, and blankets. India exported carpets worth approximately US $1.3 Bn in the year of 2024. The production of wool is concentrated in the states of Rajasthan, Jammu and Kashmir, and Gujarat. The Integrated Wool Development Programme (IWDP) by the Ministry of Textiles has initiated a comprehensive platform for small wool manufacturers, providing them access to marketplaces and linking the wool supply chain in the country.

3 Energy Consumption in the Indian Textile Industry

Energy accounts for 15–20% of the total production cost in the textile industry. About 23% energy is consumed in weaving, 34% in spinning, 38% in chemical processing, and 5% for other miscellaneous uses [8].

3.1 Energy Consumption in the Production Process

The following processes occur once the fibre has been extracted from its raw state and spun to obtain yarn.

3.1.1 Yarn Manufacturing

This foundational stage involves several sub-processes, including fiber cleaning, blending, carding, drawing, roving, and ring or rotor spinning. These operations are typically driven by electric motors, which account for a large proportion of electricity consumption. Humidity and temperature control systems are essential to maintain fiber integrity and prevent static buildup, which further increases power usage. Older machinery tends to be less energy-efficient, amplifying energy waste in traditional mills.

3.1.2 Fabric Production

Once yarn is prepared, it is converted into fabric through weaving or knitting. Weaving machines, especially air-jet and rapier looms, consume a high amount of electricity due to the mechanical force required and the use of compressed air. Knitting machines, both circular and flatbed, also operate continuously, often 24/7, in high-output facilities. Ancillary systems such as compressors, lighting, and automated transport add to the total energy load.

3.1.3 Wet Processing

This stage is widely considered the most energy- and water-intensive phase of textile manufacturing. It includes de-sizing, scouring, bleaching, dyeing, printing, and finishing. Large amounts of steam and hot water are required to heat treatment baths and ensure consistent chemical reactions. Most mills still rely on coal- or oil-fired boilers, which not only consume vast energy but also release significant pollutants. Effluent treatment plants (ETPs) that operate alongside are energy-demanding but critical for regulatory compliance.

3.1.4 Garment Manufacturing

Although considered less energy-intensive, garment making is still a significant contributor to the industry's overall energy consumption. Activities such as fabric cutting (manual and laser-based), sewing, embroidery, washing, ironing, and packaging require dependable power supplies. Sewing machine motors, vacuum tables, heat pressing systems, and conveyor belts collectively create a non-negligible energy demand. In export-oriented units, air-conditioning, lighting, and IT infrastructure further contribute to electricity usage.

Each of these stages also relies on cross-cutting support systems like:

- Compressed Air Systems: Used in looms, spinning machines, and dyeing units.
- Water Treatment and Recycling Units: Require pumps and filtration systems.
- HVAC (Heating, Ventilation, and Air Conditioning): Necessary to maintain optimal conditions for equipment and personnel, especially in spinning and garment units.
- Lighting and Control Systems: Automation and production monitoring systems add to the baseline energy load.

3.2 Sources of Conventional Energy

3.2.1 Electrical Energy

The main use of Electrical Energy is in the stages of spinning, weaving and wet processing of yarn and fibres. A Significant portion of the electrical energy is spent on operating the various machinery required for these processes. Most of the textile mills draw power from the state electricity boards but the differences in tariff rates across states have led to implications on the cost of production. As a result of mass protest, the government of India is considering a 'One Nation, One Power tariff to bring uniformity in the textile industry.

3.2.2 Thermal Energy

Thermal Energy is primarily used in the chemical processing stage, including heating and drying of water. The thermal energy is transferred as steam through pipes that are generated through fossil fuels. Around 25–30% of heat generated is lost during steam generation and distribution [9]. Therefore, alternatives for energy sources through solar, wind, tidal and other resources are being experimented.

4 Renewable Energy Alternatives

4.1 Solar Energy

About 50% of the energy used in a textile mill can be replaced with solar energy, resulting in reduced emissions and wastage. Solar is an excellent alternative to be implemented in place of thermal energy for heating and drying purposes, and can be used in the dyeing, bleaching, washing and for steam generation processes. Solar is unlimited in nature and ensures a long lifespan for the source of energy. The excess solar generated during the daytime can also be preserved through heat storage for operations in night or during cloudy/monsoon days. The energy savings from offsetting electricity with solar have proven to be significantly reducing the cost of production despite a high initial cost of installation for the solar panels. The average payback period for the investment is only 3 years as firms reap economies of scale due to the long life of panels and the solar equipment system. In India, solar has been adopted by many manufacturers in Gujarat and Tamil Nadu particularly parts of Coimbatore and Tiruppur, opting for rooftop solar systems that will reduce the line losses as the site of power generation and consumption are the same.

4.2 Wind Energy

The textile industry in Tamil Nadu is the single largest consumer of wind energy and continues to propel further installations of wind mills in the state. About 80% of the installed capacity is used for generating electricity in the textile mills and factories for various procedures such as carding, spinning and weaving. The capacities of the wind turbines vary from 225 kW to 2.1 MW, depending on the power requirement in the textile factory. Trippur knitwear cluster located in Tamil Nadu which contributes to 60% of country's knitwear exports has transitioned completely to cleaner energy sources and zero liquid discharge. The effluents from the chemical and dyeing processes are treated and reused in the industry. Its ideal location in the bridge of mountains helps in channeling wind flow for power generation. Currently it generates 1950 MW using wind turbines, of which the cluster's consumption is only 350 MV. The surplus energy is supplied to the Tamil Nadu electricity grid for use in other purposes [10]. Other potential areas for wind energy generation in India, are the northern parts of Gujarat and Rajasthan with their proximity to major textile hubs such as Ahemedabad and Surat.

4.3 Green Hydrogen

Textile industry comprises highly water-intensive manufacturing firms that also generate a significant amount of wastewater and CO_2 emissions. By integrating hydrogen generated through electrolysis, the dual problem of energy efficiency and wastewater management is addressed. Green hydrogen has been proven as a reliable alternative to electricity that can be implemented in the textile industry. This can be done in two ways:

- Hydrogen Fuel Cells: These cells generate electricity from hydrogen for the textile manufacturing process replacing traditional fossil fuels. The additional benefit of hydrogen fuel cells is that the only byproduct of generating electricity is water, which can be used further in the manufacturing process of the textiles.
- Combined Heat and Power Systems (CHP): The CHPs use hydrogen to generate electricity and heat simultaneously, reducing overall emissions.

Ongoing research in the field suggests a high possibility of converting fossil fuel-based textile plants into hydrogen fuel systems, as it is highly compatible with other renewable sources such as solar and wind energy. Leading Indian textile manufacturers are adopting Green hydrogen, integrating it into operations, transportation, and waste management uses.

4.4 Hydropower

The textile industry consumes a large portion of water in its processing, such as dyeing, printing, and finishing. The wastewater generated is highly toxic with dyes and chemical mixed effluents that have an adverse effect on the environment. Before the invention of electricity, hydropower was an ancient source of energy for the textile mills. Hydropower works by generating electricity due to the movement of water from an area of higher altitude to an area of lower altitude. Studies have shown that the wastewater ejected by the textile industry can be reused as a source of hydropower for the machinery, thereby creating a circular economy with no wastage. The implementation of hydropower plants requires the presence of flowing water near the manufacturing plant, which might restrict wide-scale implementation; however, a significant proportion of textile mills are located near waterways for easy access.

5 Decentralised Renewable Energy

Unlike centralised energy supply controlled by the state or union government, decentralised energy systems are set up by individual textile mills and apparel manufacturers, which gives them autonomy to control the power supply required for their firms. The decentralised renewable energy systems are characterised by their location closer to the production facilities as the heat generated cannot be transported over long distances. The independence from the central grid has helped significantly in reducing the cost as well as boosting the scale of production. The decentralised renewable energy sources are able to supply the required heat and power with much reduced inefficiencies as compared to the central grid.

Small manufacturers in a region can contribute to set up a decentralised source of renewable energy that can be used efficiently in the production of textiles in the area. Studies show that the excess electricity generated can be connected to the central grid to earn additional income, or kept as storage to be used during peak demand. Shifting to decentralised power generation allows for coordination between heat and power generation in CHP plants.

A survey conducted in India shows that the implementation of DRE has improved the productivity and income of small rural firms. It has also been noted that the machines improved the quality of yarn produced. DRE has been gaining popularity among the textile manufacturing units in India due to the reduced cost of production and emissions.

6 Cost Benefit Analysis

The cost-benefit analysis between conventional sources of energy, such as grid power and diesels, against renewable sources of energy, such as solar and wind, is evaluated based on the differences in capital outlay, operating cost, lifetime, return on investment (ROI) and CO_2 emissions.

- Capital Outlay—Compared to grid electricity, the initial installation cost of renewable energy is significantly higher, estimated at 4 crores in India, with a payback period of 3–4 years.
- Operating Cost—The operating cost is higher for the grid power as it requires payments to the governments and regular maintenance fees, estimated 35–40 lakhs. The solar panels have less maintenance and operating cost, unlike changing rates of the electricity grids, which affect the cost of production.
- Lifetime—The lifetime of traditional fossil fuel-based machinery would be only 10 years compared to the 25-year lifespan of solar panels with minimal maintenance.
- Emissions—The conventional energy source would release 500,000 kg of CO_2 every year, while renewable sources do not harm the environment.
- Return on Investment—Conventional sources have no real return on investment; power is a recurring variable cost dependent on usage. Renewable sources save on electricity costs. Renewable energy has a payback period of 3–4 years with a return on investment of 25% higher than a conventional energy source.

The overall CBA highlights that, while conventional energy sources are easily available and reliable, renewable energy sources are more cost efficient and climate conscious. The future holds high potential for alternative sources of energy, and early transition can help small and medium manufacturers in the textile industry to take advantage of the less exploited sector.

A detailed CBA analysis on each step of the textile manufacturing process can be used to capture the significant deviation in production between conventional and renewable sources of energy, which can aid manufacturers in deciding on the transition.

6.1 Yarn Manufacturing

6.1.1 Capital Outlay

Conventional: Minimum upfront cost for connection to the grid and diesel gensets for a medium-sized enterprise.

Renewable: Higher upfront cost for installation for a 1 MW solar wind hybrid system capable of meeting 60% of the spinning mill's electricity demand.

6.1.2 Operating Cost

Conventional: Annual operating cost is erratic and is dependent on state electricity tariffs that are not uniform across states. Diesel backup increases further costs in case of a power shutdown.

Renewable: Low maintenance cost on cleaning panels and turbines. Stable and constant prices.

6.1.3 Lifetime

Conventional: Typically 10–12 years, before replacement and significant damage due to erratic power supply and shutdowns.

Renewable: Have a longer lifespan, an average of 15–20 years, with minimal degradation.

6.1.4 Emissions

Conventional: The average electricity demand of a spinning mill is 1250–1500 kWh/ton of yarn. For a 10,000-ton yarn per year unit will generate approximately 12,300 tons of CO_2 annually [11].

Renewable: Negligible emissions. Solar wind hybrid systems reduce emissions by 10,000–12,000 tons annually on an average.

6.2 *Fabric Production*

6.2.1 Capital Outlay

Conventional: The capital outlay for conventional units typically consists of installing looms, warps and air compression systems for the weaving processes. The investment is approximately ₹1–1.2 crores.

Renewable: The energy-efficient compressors and HVAC systems for humidity control requires a higher upfront cost of ₹4–5 crores for installation [12].

6.2.2 Operating Cost

Conventional: Operating costs are significantly higher as looms and other machinery run continuously, often 24/7. These add to the energy burden.

Renewable: As solar and wind systems have lower recurring costs, the operational expenditure with renewable systems is brought down by 15–20 lakhs per year.

6.2.3 Lifetime

Conventional: Powerlooms and air compression systems dependent on grid energy have an average lifespan of 12–15 years.

Renewable: These systems last for 20–25 years without much depreciation and minimal maintenance costs.

6.2.4 Emissions

Conventional: In terms of emissions, a mid-sized weaving unit consuming grid electricity emits roughly 8000–9000 tons of CO_2 annually. This comes mainly from the electricity used in high-demand air-jet looms and compressors.

Renewable: By shifting to renewable energy, emissions can be reduced by nearly 7000 tons per year [13].

6.3 Wet Processing

6.3.1 Capital Outlay

Conventional: The capital expenditure for diesel-powered machinery, such as boilers in the dyeing and chemical processing units, is estimated to be ₹2–2.5 crores [14].

Renewable: Alternative options, such as biomass boilers and hybrid hydrogen systems, require a higher investment of ₹5–6 crores [12].

6.3.2 Operating Cost

Conventional: As conventional machinery is fuel dependent, volatility in fuel and oil prices makes operating costs unreliable estimated to ₹1–2 crores annually.

Renewable: Through the use of low cost biomass residues and solar thermal systems, the operation costs are reduced to 40–50 lakhs and are significantly cheaper based on case studies from Tiruppur, Tamil Nadu (Confederation of Indian Industry [CII], 2013).

6.3.3 Lifetime

Conventional: The lifetime of diesel boilers is 10–12 years, with frequent repair requirements.

Renewable: Biomass and other thermal boilers last up to 20 years with minimal maintenance.

6.3.4 Emissions

Conventional: A medium-sized wet processing plant emits nearly 15,000 tons of CO_2 annually [11].

Renewable: By adopting renewable sources, the emissions can be reduced by 12,000–14,000 annually, improving the quality of produce and the environment [13].

6.4 Garment Manufacturing

6.4.1 Capital Outlay

Conventional: The machinery required for garment and apparel manufacturing differs from fabric to finish. The capex is estimated at around 25–30 lakhs, which includes grid connections and backup diesel generators [12].

Renewable: The capital outlay for renewable energy is more expensive for the installation of rooftop solar to generate at least 80% of the energy [14].

6.4.2 Operating Cost

Conventional: The operating cost of traditional machinery is higher due to reliance on grid and diesel systems, estimated at around 15–20 lakhs per annum.

Renewable: Renewable energy sources significantly bring down the operating cost to 3–5 lakhs due to lower maintenance requirements.

6.4.3 Lifetime

Conventional: Traditional machinery lasts only 8–10 years before replacement and also requires several maintenance works during its lifetime.

Renewable: All renewable components last 15–20 years with minimal maintenance requirements.

6.4.4 Emissions

Conventional garment units emit around 2500–3000 tons of CO_2 annually, while renewable adoption can reduce emissions by approximately 2000 tons each year [13].

7 Barriers to Transition

Despite the growing awareness of sustainability, numerous barriers continue to hinder the widespread adoption of renewable energy in the textile industry. These obstacles range from financial and infrastructural to regulatory and educational:

High Capital Investment: One of the primary barriers is the substantial upfront investment required for renewable installations. Small and medium-sized enterprises (SMEs), which dominate the textile sector in countries like India and Bangladesh, often lack access to affordable financing. Without access to low-interest loans or favorable leasing options, many find the return on investment (ROI) period too long to justify the transition.

Technical Knowledge Gaps: The successful implementation of renewable systems requires specific technical expertise. In many regions, there is a lack of trained personnel to design, install, and maintain these systems. Additionally, plant managers and owners may not fully understand the operational requirements or long-term benefits of renewable energy technologies, leading to hesitation in adoption.

Regulatory Hurdles: Renewable energy projects often face bureaucratic delays, particularly in obtaining permits, connecting to the grid, or securing subsidies. In some states, outdated regulations or misalignment between central and state-level policies lead to confusion and delays in project execution. Furthermore, limits on net metering or capped subsidies discourage larger installations.

Space Constraints: Urban-based textile facilities, or those operating in densely populated industrial zones, may not have the physical space required for installing large-scale solar panels or wind turbines. Rooftop solar systems are limited by structural load capacity and available surface area. In such cases, off-site renewable procurement or shared community energy models are potential alternatives, though they too require regulatory clarity.

Lack of Standardization and Vendor Reliability: The renewable energy sector, particularly in developing regions, suffers from a lack of quality standards and reliable vendors. Poor system design or low-grade components can lead to inefficiencies and frequent breakdowns, deterring others from making similar investments.

Solutions to these barriers include targeted government support, enhanced public-private partnerships, third-party ownership models (such as Energy Service Companies or ESCOs), and concessional funding from climate finance institutions. By mitigating financial risk and improving trust in renewable technologies, these approaches can accelerate industry-wide transformation.

8 Conclusion

The transition to renewable energy in the Indian textile industry is not only a necessity but also of strategic importance. As one of the most energy-intensive sectors, textiles have long contributed to greenhouse gas emissions, water pollution, and high operational costs through their dependence on fossil fuels. However, this also positions the industry uniquely to showcase impactful change. On the demand side, consumers prefer products with a green label providing a competitive advantage in the industry. On the supply side, initiatives such as the DRE can significantly reduce energy costs for manufacturing firms, thereby boosting revenue and sales.

While the road to a fully renewable-powered textile industry is complex, involving high capital investment, technical upgrades, and regulatory reforms, the long term benefits are profound. Companies that have embraced renewables are already witnessing reduced energy bills, improved brand perception, enhanced market access, and alignment with international sustainability frameworks such as the Paris Agreement and the UN Sustainable Development Goals.

Strategically, renewable energy adoption makes the textile sector less vulnerable to fuel price volatility and energy supply disruptions, which are becoming increasingly common due to geopolitical conflicts and climate-related events. Economically, green energy creates new opportunities for employment, innovation, and infrastructure development. Government-backed incentives and international financing mechanisms are making the transition more feasible, particularly for small and medium-sized enterprises.

The feasibility of renewable energy adoption in India's textile industry is strongly supported by the country's geographic diversity and robust policy framework. States such as Tamil Nadu and Gujarat have become frontrunners in solar and wind integration, powering large textile clusters in Coimbatore, Tiruppur, and Ahmedabad. Punjab, Maharashtra, and Karnataka offer strong potential for biomass-based solutions, leveraging abundant agricultural residues, while Himalayan and north-eastern states present opportunities for small-scale hydropower integration into textile hubs. India's policy push further enhances this feasibility: the National Solar Mission [15] and Wind-Solar Hybrid Policy [16] have enabled industrial adoption at scale, while the National Green Hydrogen Mission [17] positions hydrogen as a transformative alternative for high-heat textile processes such as dyeing and finishing. However, despite these advantages, adoption remains uneven due to capital costs, infrastructure bottlenecks, and technological gaps. To fully unlock potential, stronger research and development (R&D) is required in areas such as advanced solar thermal systems, hydrogen-based combined heat and power (CHP), tidal energy, and energy storage solutions. Expanding the renewable portfolio beyond mainstream solar and wind will not only reduce the textile sector's carbon footprint but also ensure long-term energy security, cost stability, and competitiveness in alignment with India's Nationally Determined Contributions (NDCs) under the Paris Agreement and the UN Sustainable Development Goals (SDGs).

In essence, the textile industry's transition to renewable energy is not just a compliance-driven adjustment but a fundamental shift toward future readiness. With sustained investment, knowledge sharing, and commitment to sustainability, the sector can evolve into a global exemplar of how traditional industries can adapt and flourish within a green economy. This journey, though demanding, offers a vision of a more sustainable, equitable, and innovative textile future.

References

1. Press Information Bureau (2025) Textile sector registers significant growth… https://www.pib.gov.in/PressReleasePage.aspx?PRID=2156220
2. India Brand Equity Foundation (2025) Textiles industry in India. https://www.ibef.org/industry/textiles
3. Grand View Research (n.d.) India natural fiber market outlook. https://www.grandviewresearch.com/horizon/outlook/natural-fiber-market/india
4. Cotton Association of India (2024) Crop area & production report. https://caionline.in/news_reports/cab_area_production
5. Palamutcu S (2010) Electric energy consumption in the cotton textile processing stages. Energy 35(7):2945–2952. https://doi.org/10.1016/j.energy.2010.03.029. https://www.sciencedirect.com/science/article/abs/pii/S0360544210001489
6. Das S (n.d.) Studies on the utilization of hybrid energy in domestic silk reeling basin. [Manuscript] 2010
7. Tamil Nadu Agricultural University (2024) Jute report 2024–25. https://agritech.tnau.ac.in/ta/agri_info/pdf/Jute%20Report%20-%202024-25.pdf
8. Sharma S (2013) Energy management in textile industry. Int J Power System Oper Energy Manag 2:191–195. https://doi.org/10.47893/IJPSOEM.2013.1087
9. Apparel Impact Institute & Global Efficiency Intelligence (2025) Low-carbon thermal energy roadmap for the textile industry. https://apparelimpact.org/wp-content/uploads/2025/03/Aii-GEILow-Carbon-Thermal-Energy-Roadmap-for-the-Textile-Industry.pdf
10. Fibre2Fashion News Desk (2025) Tiruppur gears up for next phase of growth journey. https://www.fibre2fashion.com/news/textile-news/tiruppur-gears-up-for-next-phase-of-growth-journey-301282-newsdetails.htm
11. U.S. Department of Energy (2010) Energy-efficiency improvement opportunities for the textile industry. U.S. DOE, Washington, DC
12. Ministry of New and Renewable Energy (2023) Annual report 2022–23. Government of India. https://cdnbbsr.s3waas.gov.in/s3716e1b8c6cd17b771da77391355749f3/uploads/2024/10/20241014958242917.pdf
13. Farhana K et al (2022) Energy consumption, environmental impact, and implementation of renewable energy resources in global textile industries: an overview towards circularity and sustainability. Mater Circ Econ 4(1) https://www.researchgate.net/publication/359574045_Energy_consumption_environmental_impact_and_implementation_of_renewable_energy_resources_in_global_textile_industries_an_overview_towards_circularity_and_sustainability
14. United Nations Industrial Development Organization (2022) Industrial development report 2022: Overview. UNIDO. https://www.unido.org/sites/default/files/unido-publications/2023-03/IDR-2022-OVERVIEW-en.pdf
15. National Solar Mission, Ministry of New and Renewable Energy (2010) Towards building solar India (Mission document). https://www.seci.co.in/upload/static/files/mission_document_JNNSM(1).pdf
16. Wind-Solar Hybrid Policy. Andhra Pradesh Electricity Regulatory Commission. (2023). https://aperc.gov.in/admin/upload/APTextiles3.pdf
17. National Green Hydrogen Mission: Key highlights. Press Information Bureau. (2023). https://static.pib.gov.in/WriteReadData/specificdocs/documents/2023/jan/doc2023110150801.pdf

MIX
Papier aus verantwortungsvollen Quellen
Paper from responsible sources
FSC® C105338

If you have any concerns about our products,
you can contact us on
ProductSafety@springernature.com

In case Publisher is established outside the EU,
the EU authorized representative is:
Springer Nature Customer Service Center GmbH
Europaplatz 3, 69115 Heidelberg, Germany

Printed by Libri Plureos GmbH
in Hamburg, Germany